Mechanization of Cultivated Crops

Mechanization of Cultivated Crops

Editor

Cornelio Losa

scitus
academics

Mechanization of Cultivated Crops

Edited by **Cornelio Losa**

ISBN: 978-1-68117-542-3
Library of Congress Control Number: 2016942942

SCITUS Academics LLC,
www.scitusacademics.com
Box No. 4766, 616 Corporate Way,
Suite 2, Valley Cottage,
NY 10989

Notice

Reasonable efforts have been made to publish reliable data and views articulated in the chapters are those of the individual contributors, and not necessarily those of the editors or publishers. Editors or publishers are not responsible for the accuracy of the information in the published chapters or consequences of their use. The publisher believes no responsibility for any damage or grievance to the persons or property arising out of the use of any materials, instructions, methods or thoughts in the book. The editors and the publisher have attempted to trace the copyright holders of all material reproduced in this publication and apologize to copyright holders if permission has not been obtained. If any copyright holder has not been acknowledged, please write to us so we may rectify.

Printed in United States of America on Acid Free Paper

Preface

Cultivation is the act of looking after or possibly raising plants. Cultivation loosens and breaks (burns) the dirt. The dirt around existing plant life is actually grown (by a hoe or perhaps by a cultivator's machine) to destroy weeds and also promote growth by increasing water infiltration and aeration of the soil. The soil is ready to plant a crop and is grown by a harrow or possibly by a grower. A crop is any algae, fungus or plant grown for food, medicine, biofuel, livestock forage, clothing and several other uses. In comparison, animals raised by humans are called animals, with the exception of animals kept as animals. Mechanization helps farmers to achieve relevance in farming operations with self-esteem and to use costly inputs with reduced quantities to increase efficiency and efficiency. Marginal and small-scale farmers can now use customized high-capacity agricultural machines. The consequences of development can be observed in several phases, such as the increase in production and productivity per worker, the accuracy of the application of crop inputs, the increase in crop intensity due to the timeliness of operations, the increase in product quality, the decrease in grain losses and the increase in agricultural employment. Mechanization also encourages better management of agricultural inputs, improved work performance and employment conditions that would normally be difficult by hand. It can also help reduce production costs.

Mechanization of Cultivated Crops provides a comprehensive overview of farm tools, equipment and methods used in different operations such as land development, tillage, seed/planting, interculture, application of fertilizers, plant protection and harvesting.

Table of Contents

Chapter 1

Cultivation Techniques

Caterina Briccoli Bati[1], Elena Santilli[1], Pietro Toscano[1] and Ilaria Guagliardi[2]

[1] Agricultural Research Council - Olive Growing and Oil Industry Research Centre, Rende (CS), Italy

[2] National Research Council - Institute for Agricultural and Forest Systems in the Mediterranean, (ISAFOM), Rende (CS), Italy

1. INTRODUCTION

In an agricultural context, olive growing is emerging as a dynamic and interesting topic. The majority of Italian olive growing still requires organic renewal interventions such as farm restructuring and tree planting, varietal conversion, mechanization, technical assistance for the implementation of technological innovations, better organization and contractual weight in product marketing. Therefore, the main objective in this area can be obtained with a more economical management of olive orchards in order to achieve a high production per unit area at lower costs and while respecting the environment. Therefore, current olive growing must be based on two pillars such as the reduction of management costs and the use of cultivation techniques with a low environmental impact. More and more attention is being paid, by the EU, and also by Italy, to environmental sustainability, biodiversity and compliance requirements in agriculture. Indeed, in recent years the EU has issued a set of regulations aimed at environmental protection and enhancement of rural areas by improving the competitiveness of the agricultural sector in order to obtain high-quality products aimed at enhancing the peculiarities of the different territories of origin (PGI, PDO) and protecting agricultural and natural resources. Quality must always be considered a key resource for agriculture, which will enable farms to survive and compete in both Italian and International markets. The cultivation techniques used in olive orchards are directed to preserve and improve the physical and chemical characteristics of the soil (soil preparation and tillage, irrigation, fertilization) and to enhance plant production (training, pruning, fruiting, production and pesticide treatments). The knowledge of olive morphology and biology is a prerequisite for the rationalization of cultivation techniques to improve the quantity and quality of production.

Although the olive tree can be considered a hardy plant and is cultivated in marginal areas, it requires specific cultivation techniques coordinated and integrated with each other in order to exalt their productive potentialities. This is why the wise use of tools such as pruning, irrigation, nutrition and soil management plays an important role in achieving a greater vegetative and reproductive plant balance, and cost containment should be the main goal that guides management decisions.

2. SOIL

Soil is defined as the top layer of the earth's crust. It is formed by mineral particles, organic matter, water, air and living organisms. It is in fact an extremely complex, variable and living medium, and represents a non-renewable resource which performs many vital functions: food and other biomass production, storage, filtration and transformation of many substances including water, carbon, nitrogen. As long as 100 years ago, Wollny (1898) described the positive effect of soil structure on root growth, water availability, gas transport in soils as well as the positive effects of soil structure on soil strength. He mentioned that the mechanisms involved in the interaction between soil structure and plant growth and yield needed to be investigated. Since then, the positive effects of a favorable soil structure and the negative effects of, for example, soil compaction on crop growth and/or yield have been repeatedly described (e.g. Blank, 1932-1939; Dexter, 1988; Hakansson et al., 1988; Kay, 1990). The anthropogenic activities such as tillage, mineral fertilization, waste disposal and industrial pollution, affect both chemical and physical natural soil properties (Kabata-Pendias & Mukherjee, 2007).

Recent improvements and new methods in analytical chemistry and increasing areas of environmental investigation have substantially added to our knowledge of agricultural soil science. For example, the soil characteristics of an olive plantation are especially important in terms of vulnerability to erosion and, to a lesser extent, to leaching of potentially contaminating elements contained in fertilisers and pesticides. The root system of the olive is concentrated in the top 50-70 cm of soil although it may send out roots to a depth of more than one meter to satisfy its water needs. Therefore, the soil must have an optimal texture, structure and composition to a depth of at least one meter. The management of a cropping system requires periodic evaluation that includes systematic testing, with the aim of determining the nutritional status of soils in order to assess the existence of any nutritional deficiency, excess or imbalance and form a basis for planning the nutrient supply as well as other practices (tillage, amendment, correction). The following is a brief description of the main chemical and physical soil properties.

2.1. Soil Texture, Porosity and Density

The textural class is the first parameter that defines soil properties, and is determined by the relative percentage of the three major soil compounds: sand, silt and clay, defined by the respective particle diameter size.

Clayey grounds are characterized by particles of a diameter of less than 0.002 mm, constituted by flinty minerals with different capacities to inflate in the presence of water and to contract in dry conditions, forming cracks which are typical of vertisols. Clayey soils show a low water permeability and high plasticity, which can induce stagnation phenomena and root asphyxia in wet conditions; while in the dry state it has notable tenacity and cohesion.

Silty soils have elementary particles of greater dimensions than the clayey ones and, unlike these, they have greater difficulty in reaching a glomerular structure. The lower structural stability of silty soils causes a low macroporosity and a great bulk density, that determine conditions of low aeration, low permeability and water stagnation in the profile. This type of soil is subject to loosening conditions with greater facility in comparison to clayey soils. However, they have a greater tendency to pulverization in the dry state, and the formation of mud in the wet state.

Sandy soils have particles of a diameter between 2 and 0.02 mm; they are characterized by high permeability of rainwater, and fast mineralization of organic matter.

The Olive tree responds best to soil textures with balanced proportions of sand, silt and clay. Soils that are primarily sandy do not have good nutrient or water-holding capacities, but they do provide good aeration and olives do well, especially when water is available and the crop is properly fertilized to satisfy its mineral requirements. The soil should not contain too much clay to avoid limiting air circulation and to prevent soil management problems. The soil particles should aggregate in granules or crumbs to make the soil porous; this is ensured by sufficient quantities of organic matter and rational soil management to prevent compacting and erosion. Soil is composed of solid particles (mineral and organic matter) of different sizes, usually bound together into aggregates by organic matter, mineral oxides, and charged clay particles. The number and size of pores vary considerably among soils exhibiting different organic matter content, texture and structure and cultivation techniques have a great effect on bulk density and porosity: any management practice that increases organic matter will increase the granular structure of the soil, increase the pore space, and decrease the bulk density (Gisotti, 1988; Giordano, 1999; Hao et al., 2008).

2.2. Organic Matter and Nitrogen Ratio

The Organic Matter (OM) is a complex mixture of organic compounds deriving from metabolic wastes and decaying residuals of plants, animals and microorganisms, at different stages of decomposition. The OM percentage directly influences the structure and chemical-physic properties of soil in terms of water infiltration and retention, element absorption, particle aggregations, Cation Exchange Capacity (Al^{3+}, Fe^{3+}, Ca^{2+}, Mg^{2+}, NH_4^+), buffering power, over the nutrient source for the plant. The quantity and nature of OM is highly dependent upon farming practices and climatic conditions and is found as both chemically stable humus (or passive OM) and partially decomposed plants, microbes and animal residues (or active OM).

Measures to increase the organic content are a very important part of good soil management in Mediterranean regions, especially in order to reduce vulnerability to erosion (European Soil Bureau, 1999). Practical measures are based on the incorporation of organic matter such as farm-yard manure, cover crops, pruning and processing residues, and soil tillage.

Plant availability of organic N is dependent on OM breakdown, which is difficult to estimate. The ratio of total organic carbon and total nitrogen (C/N) is the traditional guide to the nature of the organic matter present in the soil.

The basic premise behind this ratio is that organic carbon is the primary source of energy for soil microbes, but these also require nitrogen to multiply and utilise this energy. The microbes utilise soil carbon via respiration, with the consequent loss of carbon dioxide from the soil. As the active fraction of the OM is thus degraded, the C/N ratio drops until a steady state (the passive fraction) is finally attained. Interpreting this ratio is complicated, as it also depends on the nature of the OM. The passive fraction of the OM can have a C/N ratio that is 'medium'. Consequently, medium C/N ratio soils can have a wide variation in mineralisable N status, and this is a limitation when considering the C/N ratio in isolation.

2.3. Cation Exchange Capacity (Cec), Ph, Electrical And Hydraulic Conductivity, Water Content

Plant nutrients usually exist as ions which carry an electrostatic charge. This electrostatic charge is a result of atomic substitution in the lattices of soil minerals and because of hydrolysis reactions on the broken edges of the lattices and the surface of oxides, hydroxides, hydrous oxides and organic matter (Hendershot et al., 2008a). These charges attract counterions (exchangeable ions) and form the exchange complex. Ions can be bound to the soil in varying degrees. At one extreme, they may be an integral part of the soil, strongly bound to silica and essentially unavailable to growing plants. At the other extreme, they may be fully soluble and not interact with the soil to any significant extent. Exchangeable ions are between these two extremes, and are weakly bound to soil particles. The bonds between soil particles and exchangeable ions are not permanent, and are continually broken reformed, as the ions move within the water surrounding soil particles. The bonding of these ions largely prevents their loss by leaching, but is not so strong that plants cannot extract them from the soil. In fact, plant roots absorb exchangeable ions by 'swapping' them for hydrogen cations (H^+).

The cation exchange capacity is often estimated by summing the major exchangeable cations (K, Ca, Mg, and Na) using units of cmol kg^{-1}, even if the common expression for CEC is in terms of milliequivalents per 100 grams (meq/100g) of soil. The CEC of soil can range from less than 5 to 35 meq/100g for agricultural type soils, and is related to clay and organic matter content.

CEC is important for maintaining adequate quantities of plant available calcium, magnesium, sodium and potassium in soils. For many crops the magnesium level should ideally be twice as much as that of potassium. When magnesium is lower than potassium, suppression of magnesium uptake can occur. Sodium is only of secondary importance in the soil test as its uptake by plants is largely dependent on the plant species involved and the potassium status of the soil, rather than the level of sodium extractable from the soil.

The Total Base Saturation is related to CEC, which represents the proportion of the soil's total capacity for cations that is actually occupied by these nutrients. It is calculated by summing together the levels of calcium, magnesium, potassium and sodium found in the soil and expressing this sum as a percentage of the CEC value.

Soil pH is one of the most common and important measurements in standard soil analyses (Hendershot et al., 2008b). The pH value expresses degree of acidity or alkalinity of the soil. It is important because it influences the chemical and physiological processes in the soil, and the availability of nutrients. Availability changes differently with pH levels: aluminium, copper, iron, manganese and zinc increase when the pH decreases; unlike magnesium that decreases when the pH decreases (Belsito et al., 1988;AA.VV., 1989; Jones, 2003).

Electrical conductivity (EC) is the ability of a material to conduct an electrical current and is commonly expressed in units of micro Siemens per meter (µS cm^{-1}). It is used to estimate the level of soluble salts. The measurement of EC in the soil water extracted from the field-water content is theoretically the best measure of salinity as it indicates the actual salinity level experienced by the plant root (Miller & Curtin, 2008). However, this measurement has not been widely used because it varies as soil-water content changes over time and so it is not a single-valued parameter. A soil is considered saline if the EC of the saturation extract exceeds 4000 μS cm^{-1} at 25°C. The soil EC varies depending on the amount of moisture held by soil particles. Consequently, the EC correlates strongly to soil particle size and texture and affects crop productivity.

Soil water analyses can be organized into two main groups: analysis of storage properties and analysis of hydraulic properties. The water content of soil is part of the analysis of storage properties which refer to the soil's ability to absorb and hold water. Instead hydraulic conductivity is a hydraulic property which refers to the soil's ability to transmit or conduct water. It is more difficult for plants to absorb nutrient elements at low soil moisture levels, so nutrient element contents will be lower.

3. SOIL MANAGEMENT

In order for plants to live, two key functions can be attributed to soil: habitability and nutrition. The function of habitability mainly depends on the physico-chemical characteristics of the soil. The function of nutrition depends on the factors that make nutrients bio-available to the plants, described above, determining the fertility of the soil as productive attitude. Biological soil functions depend on the micro-organic pattern, responsible for processes on the organic matter such as: mineralization, humification, nitrification, nitrogen fixation, symbioses, and parasitism. The agricultural management systems of soils, such as crop rotation, nutrient application, plant species, kind of tillage, and use of pesticides may have a strong impact on the composition of the soil microbial community. Maintenance of sustainable soil fertility depends greatly on the ability to harness the benefits of rhizosphere microorganisms such as arbuscular mycorrhizal fungi (AMF), which form a symbiotic association with the roots of most plant families. Olive plants are known to form arbuscular mycorrhiza (Roldán-Fajardo & Barea, 1985; Briccoli Bati et al., 1992; Calvente et al., 2004), the most common mycorrhizal type involved in normal cropping systems, being considered as a key component in environmentally friendly agro-biotechnologies (Jeffries & Barea, 2001).

Mycorrhizae act as biofertilizers, bioregulators, and biocontrol agents (Lovato et al., 1996; Von, 1997). Arbuscular mycorrhizal fungi allow the plant to absorb greater quantities of water and nutrients, particularly those less mobile in the soil such as phosphorus. In addition to phosphorus, other elements such as nitrogen, zinc, calcium and sulfur are involved in the mechanisms of mobilization and uptake by mycorrhizal fungi.

Mycorrhizal symbiosis also acts as a bio-regulator, able to influence some physiological processes, growth regulators and the development of the plant, to modify the morphology of the roots, the roots/foliage ratio and sometimes branching foliage and flowering. These soil fungi improve the agronomic fertility of the soil through the formation and stabilization of particle aggregates, in particular in land lacking structure.

3.1. Tillage

Tillage consists of some mechanical operations, performed with different tools, which modify soil structure, according to the management needs, that can be summarized as: increase of the soil mass (active layer); increase in soil permeability, runoff and erosive phenomena; accumulation of water reserves; reduction of evaporation due to interruption of superficial capillarity; destruction or containment of weeds; burial of fertilizers, corrective, amendants, and crops residual.

Usually, for olive groves a deeper autumnal tillage is carried out, to increase the water reserve and to bury the phospho-potassic fertilizers; while during the spring-summer period, some harrowing is performed to reduce evaporation and to eliminate weeds. Soil tillage was classified according to the epoch and the type of performance, distinguishing into preparatory practices, performed before the plantation, to constitute suitable conditions to sustain the crop after the implant, and subsequent practices, performed during crop culture.

Soil tillage can be performed with different tools classified into three main groups: mouldboards, rippers and scramblers. Mouldboards cut and upset the soil; rippers cut the profile producing clods but without modifying soil stratigraphy; and scramblers break up and remix the worked layer.

The choice of the best tillage technique must be performed in order to: reduce costs, in terms of working times and fuel needs; increase the timeliness of intervention; maintain a suitable productive level of crops and soil fertility; contain erosive phenomena.

The soil water content strongly influences the choice of the epoch and the type of intervention performance. Dependent on this, there are different physical soil states: i) cohesive, when the soil is dry; in this state the soil does not stick to utensils, it has maximum tenacity, it is crushproof and resistant to breakup; ii) plastic, characterized by a progressive warping and stickiness, that increases with an increase in damp; iii) liquid, when the soil behaves like a suspension. Tillage is difficult and harmful to the soil when the soil is sticky; over this limit, the passing of machines provokes undoing of the structural aggregates. With dry soils the work needs higher powers, and forms compact clods of varying dimensions, according to the type of performance: plowing causes large clods, while with rotary hoeing a notable pulverization and formation of small clods occurs.

A soil is considered as loosening when it has an optimal damp for the execution of tillage, approximately corresponding to half of their field capacity, with a more or less ampler range depending on the soil type and intervention (Bonciarelli, 1981; Giardini, 1986).

Plowing is the most known and commonly used form of soil tillage in agricultural practice and is performed with three type of tools: the ploughshare plow and the disk plow, that work by traction; and the rollover plow, that acts by the tractor power take-off (PTO).

The plow operates by cutting and overthrowing a slice of soil, with an angle varying with the type of plough, the operating velocity and the operation goal: the complete overthrow of slices is necessary in green manuring and in weed control; while vertical slices improve airing and rainwater infiltration. Using a cylindrical bending breast a greater crumbling action is achieved; while a helical breast favours the slice overthrow with smaller production of thin soil. The speed of ploughing acts on both the slice overthrow and crumbling: a fast ploughing enhances the inversion of layers and the pulverization of clods. The ploughshare action can cause compaction of the deep soil, called tillage pan. Such a drawback can be enhanced using tractors working "within furrow", and in conditions of high damp. The tillage pan hinders the vertical movement of rainwater and the gaseous exchanges in the soil and the growth of the roots.

The drawbacks of the ploughshare plow are: excessive clod level, that requires other refining tillage, with further passages of machinery that stamp on the soil and degrade its structure; formation of tillage pan; high requirement of traction power. Such negative aspects can be mitigated using the disk plow, in which the ploughshare and the breast are replaced by a spherical cap, free rotating on an axle angled in respect of the operating direction. The disk limits attrition and needs of traction power. During rotation the cap lifts a slice of soil that is then crumbled and remixed. In comparison to the classical plow it better prepares the bed for seeding; it is proper for light ploughing in loose or medium textured soil rich of skeleton and in groves.

There are other tillage techniques that can be carried out, using different types of machinery which can be complementary or alternatives to ploughing. Among these the most common are:

- Ripping technique is characterized by the vertical breakup of the soil without inversion of the layers. The tools used in this type of tillage are constituted by a varying number of anchors (from 1 to 5) that practice a different action on the soil depending on their shape and interaxle. Some machineries are endowed with vibrating tools that enhance shattering of the soil, and can be joined to rolls or harrows to finish up and to level the surface in a single pass.

- Rotary Hoeing consists in the shattering and remixing of the soil performed by machinery moved with a tractor PTO, with tools that work on a horizontal axle (millers) or vertical axle (rototiller). Their drawback is the excessive shattering of soil, that worsens its structure, with compaction, formation of superficial crust and tillage pan, fast tool usury; while requiring high working power. They are unsuitable in heavy and/or skeleton soils.

- Weeding is a tillage practice complementary to ploughing, carried out to reduce the clod of the soil, and to bring up weed roots. This machinery is formed by bent rigid or elastic rippers and with different types of feet.

- Harrowing is performed to refine clods, eliminate weeds, bury fertilizers and break the superficial crust. The tools for this operation are of different shapes and dimensions, according to soil types, soil conditions and the needs of soil refining.

All tillage up to 15 cm of depth is included in the minimum tillage group,, with the aim of energy saving, preserving soil structure and timeliness of work. Among the different operative options, various types of machinery are available that can perform tillage, fertilization, seeding and chemical weeding of the soil in a single pass (Toscano, 1998).

3.2. Tillage: Soil Characteristics and Erosion

The usable kind of tillage depends on the soil texture. In clayey soils, the minimum tillage can have positive effects on the containment of erosive phenomena, due to the residual crop on the soil surface, and on compaction, in order to reduce passing of the machinery.

In silty soils all the tilling techniques that do not involve the inversion of layers favour soil structuring, and the presence of residual crop, and avoids the destructuring caused by the beating action of rain water.

In sandy soils, the choice of tillage techniques should exclude deep intervention, while all the minimum tillage techniques generally guarantee best results.

The different handling of the soil can determine a different availability of the nutrient elements, as well as a different biological activity. The techniques that do not involve the inversion of layers, allow maintaining or increasing the organic matter in the soil. With regards to the availability of the principal nutrient elements, the effects produced by the different tilling techniques vary according to the different movement of each element in the soil.

Nitrogen results mostly available for plants in a worked soil, due to the high aeration of the mass and for the velocity whereby the residual and organic fertilizers are degraded into mineral elements which can be assimilated.

For the phosphorus, a low mobile element, there are strong differences in its stratification along the profile according to the type of tillage. With a plow this element has the tendency to distribute itself in a more homogeneous way in the soil in comparison to how it is distributed in soil that has not been worked, or only cracked soil, in which phosphorus remains in the most superficial layers.

The common good supply of natural Italian agricultural land, particularly clayey ones, leads to a substantial independence of availability of potassium from the plowing technique adopted.

All tilling techniques that improve soil permeability, and allow maintenance of a vegetable coverage, are very useful in the control of erosive phenomena (Stein et al., 1986; Rasiah & Kay, 1995; Raglione et al., 2000; Toscano, 2000; Toscano et al., 2004a). Erosion consists in the removal of the most fertile soil layers by wind and/or rainwater action. The eroded amount is proportional to the intensity of the rainfall, to the slope and type of the soil.

3.3. Soil Grassing

To protect the soil profile, structure and edaphic biocenosis, it is useful to apply less expensive cultivation techniques which have a lower environmental impact than traditional tillage. The minimum tillage or controlled soil grassing, generally determine a great activity of the soil biota, due to the greater presence of organic matter and the low trouble of soil.

It is possible to implement soil grassing, which can be either natural or artificial, and to partially or wholly cover the orchard surface. The benefits of controlled grassing in olive orchards are: improvement of the soil structure, increase of soil organic matter and water absorption, reduction of runoff and erosion, improvement of carrying capacity and reduction of compactness, enhancement of microbial activity and nutritional balance; simplified management at lower operating costs. The possible competition of turf for water and nutrients with regards to the olive plants (Pastor, 1989), can be prevented with proper grass management, such as cutting, or additional fertilization (Toscano et al., 2004b).

The simpler type of soil grassing is "permanent", whereby the soil is constantly covered by spontaneous vegetation that is periodically mowed or shredded.

Alternatively, in dry summer conditions, temporary grassing can be adopted, eliminating grass, when competition for water competition begins, with superficial tillage or using contact herbicides; the coverage naturally reestablishes itself with the resumption of the rainy season, retaining its beneficial effects up to the following spring.

For artificial grassing the choice of the essences is very important, these must have fast growth following planting and to be resistant to pounding and to mulching. However, the artificial grassing presents some economic and managerial limits such as the difficult choice of the essences and the seeding costs.

The machinery for grass management consists in the rotary mower, and the shredder. They are of great working capacity and have low power needs, compared to the tools that operate on the soil; the shredder has the advantage that it grinds the mown grass, thus reducing degradation times, and it can also be used for pruning residues. Both these machines can be equipped with intercept rotary mower, which allows cutting of the grass along the row, avoiding damage to the tree trunks. Long-time experiences of controlled grassing in different non-irrigated olive orchard soils have confirmed the effectiveness of this technique in improving soil properties, in the drainage system, in the control of soil erosion and on olive tree productivity (Briccoli Bati et al., 2002; Toscano et al., 1999, 2006). On the contrary other tests, comparing different application methods of grassing, have evidenced better agronomic and productive results with green manure in summer, to avoid the increased competition for soil water occurred by permanent grassing in this environment (Toscano, 2009). Therefore, the choice of the best soil management system must be made according to the specific soils and crop characteristics. In olive groves the replacement of tillage with other techniques is possible, according to water availability, in order to obtain the best maintenance of soil fertility, a reduction of the erosion in slopes, a timeliness of intervention, a reduction of the costs and, therefore, the attainment of greater incomes.

4. NUTRITION

The olive tree is still often considered a rustic plant, having little nutritional requirements and capable to survive even in rough environments, with minimal care and management. The olive plant grows in most soil types as long as they are well drained. These plants, could also vegetate in the absence of fertilization, but require suitable nutrition to express their productive and qualitative potentialities.

In the traditional olive-grove plant nutrition is mainly based on systematic and massive inputs of chemical nutrients distributed to the soil, not always correctly and often unnecessarily (if not harmful) for plants and the environment (groundwater pollution). In many cases olive tree fertilization is often empirically approached and farmers apply much more fertiliser than the crop really needs (Tombesi et al., 1996).

The compilation of adequate fertilization programs, in terms of type, doses, epochs and disposal of the nourishing elements, are not of simple generalization and depend on local environmental and climatic factors, as well as on the effectiveness of the fertilizer composition and its application method.

It is indispensable to carry out a prior analysis of soil chemistry and of the nutrient contents of the plant by plant tissue analysis (usually leaves are used). These analyses will give significant data on the status of both soil and plant, indicating the most useful typology and doses of nutrients to apply in the fertilization plans.

Leaf analysis, is a reliable method for assessing the nutritional status of the crop (Bouat, 1968;Freeman et al., 1994). The content of the major nutrients in the leaves differs not only according to the cultivars, the soil and climatic conditions of the cultivation area, the time of sampling for the analysis but also in relation to the pruning and irrigation applied to olive orchard (Briccoli Bati et al., 1995). Some research on a regional scale has defined certain relationships between the time of leaf sampling, the foliar nutrient content and the quality of production (Failla et al., 1997; Soyergin et al., 2000). In fact, it was found that the leaf diagnostic at flowering is conclusive for the less mobile elements (Ca, Mg, Fe and Zn) while during the winter rest foliar analysis better shows the nutritional potential of the soil for nutrients with increased mobility as N, P and K (Failla et al., 1997). The level of global nutrition, generalizing, expressed as percentage amount on the leaf dry matter for nitrogen, phosphorus and potassium, results as 3,5% divided, respectively, in 2,1 - 0,35 - 1,05 with a physiological relationship of 6:1:3.

In profitable olive growing, the nutritional needs also vary in relation to the phenological phases, to the climate, to the cultivar, to the trees' productive potentiality, and to the olive orchard management, i.e. presence of soil grassing and irrigation. For these reasons, fertilization planning cannot be approached as a standard procedure and many authors report different evaluations about the nourishment needs of olive trees (Natali, 1993; Petruccioli & Parlati, 1983).

Fertilisation systems include: chemical fertilisers (NPK applied beneath the tree canopy projection, usually in the form of combined fertilisers), organic fertilisation (green and animal manures, leaves, compost, manufactured organic fertilisers), and fertilisation through watering systems and through foliage.

During the first three years of the olive plantation, when vegetative activity prevails on fructification, it is important to stimulate, with fertilization, rapid canopy and root growth of the tree to predispose the plants quickly to flowering and fruiting (Palese et al., 1997). In this phase, Nitrogen is the essential element, while phospho-potassic fertilizers at this time are less important, provided that during the preparatory work of the soil for planting, such fertilizers were distributed over the entire surface and buried with deep tillage.

When the plant completed the first phase of growth (5th - 6th year) and during the entire life of the orchard, the scope of fertilization is to induce and support the yield and, simultaneously, also to ensure the renewal of fruiting shoots and roots.

In order to calculate the amount of nutrient supply to plants it is helpful to adopt the returning criteria of nutrients removed with fruit harvesting, with pruned wood and abscised leaves: for 100 kg of drupes produced the olive tree needs around 900 g nitrogen, 200 g of phosphorus and 1.000 g of potassium. In fertilization planning, such doses must be triplicate, due to the losses leaching, volatilization, fixation, etc.

Traditionally nitrogen is supplied annually and divided in at least two doses. Most of the quantity to be given (2/3) at the end of winter before flower bud differentiation and before the growth of new lateral shoots, and the second during the flowering period (from the pre-flowering stage till fruit set). Usually the recommended nitrogen application ranges between 500-1500g for bearing tree, according to canopy volume.

Throughout the life of the olive-grove, phosphorus and potassium supply must be repeated every 5-6 years, with the doses defined by the results of soil and leaf analysis. These fertilizers are usually supplied in autumn, and burying with shallow tillage, on alternate inter-rows to limit damage to the roots, with doses of 200-400 units of potassium and 100-200 of phosphorus per hectare integrated with suitable doses of organic matter (manure, green manure or compost).

During the annual cycle, nitrogen absorption is more intense from the flowering up to the pit hardening, while the contents of N and P decrease in the leaves up to the pit hardening and at the same time they increase in the drupes. Subsequently, both in the leaves and fruits nitrogen and phosphorus decrease after veraison. Instead, potassium, constantly decreases in the leaves, while increasing in the fruits.

Biennial or triennial interventions for phosphorus and potassium, in the poor soils are useful, applied after harvesting in concomitance with deep tillage for rainwater storage, in old or dry-condition raised olive orchard; or at the end of winter, with lighter tillage in a young and intensive olive grove.

4.1. Nutrients Typology and Effects

The use of the appropriate fertilizer at the right time increases the efficiency and reduces the cost of fertilization, with a positive impact on the produced olive fruit and oil content.

Fertilization can be distinguished into organic and mineral. The first one has the purpose to improve the physical characteristics of the soil, such as the structure, the porosity, the permeability, the tackiness, the consistency, the water retention, and the pH. The second one is destined to feed the plants.

Nitrogen is fundamental in plant growth, it participates in the formation of amino acids and in the formation of proteins, therefore, it is crucial in the growing processes when the plant is young. In adult trees nitrogen supports the formation of shoots, a necessary condition to ensure constant productivity and positively influence flower formation, fruit setting and fruit development, especially during the early stages, up to the pit hardening. Nitrogen fertilization consistently increases the olive yield but only when leaf N is below the sufficiency threshold (Hartmann,1958). It is usually applied to the soil using urea, ammonium sulphate, or ammonium nitrate. Nitrogen can also be supplied with either organic materials such as feathers or blood meal, compost, or a leguminous cover crop. Its deficiency is manifested by decreased growth activity, leaf yellowing, high ovary abortion, low yield and alternate bearing (Cimato et al., 1990; 1995).

Phosphorus also has a role in growth, being essential for cell division and the development of the meristematic tissues, fruit set, fruit growth and maturation, and in lignification of the shoots. Even if absorbed by the olive tree in a relatively high quantity, the effects of phosphatic fertilization are nevertheless manifested with extreme slowness. The most used phosphate fertilizers are soluble phosphates and in particular superphosphates with 35-45% of phosphorus pentoxide, which is the form absorbed by the plant. A content of 50-100 ppm of phosphorus pentoxide in the soil detected by soil analysis, can be considered to be satisfactory. The symptoms of deficiency of this element, which is very rare, occur with a reddish or purplish coloration of the green parts of the plant, and it causes metabolic problems for growth and fructification, with delays in drupe maturation.

Potassium promotes the accumulation of carbon hydrates such as starch, an energetic reserve for metabolic processes. This element, regulates the water consumption of the plant through an increase in water retention in the tissues and it also controls transpiration. Potassium is an enzymatic activator, it increases the plants resistance to thermal extremes and to some fungal diseases, and it promotes oil accumulation in the fruits. This element is, usually, applied to the soil during winter in order to gradually reach the rooting zone with the rain. Regular potassium fertilization is necessary in order to maximize both yield and quality, especially in heavy yield years.

After nitrogen, phosphorus and potassium, other very important nourishing elements are magnesium and calcium. Magnesium is an essential component of chlorophyll and generally it is not considered in fertilization plans because it is already contained in many fertilizers. Occasionally, magnesium deficiency can be revealed in orchards growing on sandy, neutral soil. Fertilization based on magnesium sulfate corrects this deficiency.

Calcium is vital to olive plant growth, because it is an essential constituent of cell walls and contributes to the mechanical resistance of tissues, it also acts as an activator of some enzymes. Deficiencies of calcium due to soil acidity, can be corrected with an adequate lime supply as calcium carbonate.

Sulphur is present in plant amino-acids such as cystine, cysteine and methionine and is located in the soil in the organic matter. Fertilizers containing sulphur as ammonium or potassium sulphate, etc. are distributed against possible deficiencies of this element.

The most important microelements are iron, copper, zinc, manganese, molybdenum and especially boron, all developing a specific and exclusive role as enzymatic activators in the biochemical processes of the plants. These elements, present in small amounts in olive tissues, have a very narrow range between a sufficiency and toxicity level. Leaf tissue analyses provide excellent information in order to directly diagnose the toxicity or the lack of these microelements.

Above all it is very important to know the boron content of the leaves because it plays a major role in pollen growth, fruit set and plant productivity. Visible symptoms of boron deficiency are manifested with leaves with apical chlorosis, followed by necrosis and leaf drop. In the cases of a slight boron deficiency, the fertility of the flowers is reduced due to increased ovary abortion (Perica et al., 2001). Boron deficiency is nevertheless removable with extreme rapidity and effectiveness through leaf treatments during the pre-flowering stages. Foliar applications have had statistically significant effects on the yield and leaf B contents, therefore, the most economic dose was found to be 0.4% foliar application of sodium tetraborate.

The organic matter in soil plays a central role in controlling the availability of N, P and K and it can also act as a chelate, making certain micronutrients more available for the roots in the form of complexes.

4.2. Fertilization Techniques

Plant nutrition is physiologically dependent on the absorption of nourishing elements through the roots; it is therefore necessary to ensure that in the active soil layer there is a suitable endowment of available nourishing elements for the plants. Normally fertilizers are spread on the soil. Nitrogenous fertilizers, nitric, ureic or ammoniacal, are used annually according to requirements and the time of intervention, the first one being easily soluble with a fast effect, while ureic and ammoniacal products have a longer acting time and greater persistence. The nitrogen amount usually provided is of 1kg N equal to, approximately, 5kg of ammonium sulfate, 3kg of ammonium nitrate, 4kg of calcium nitrate or 2kg of urea. The principal provisions of phosphorus and potassium, due to their low mobility in the soil, are provided before the plantation of the orchard establishment, further applications are made every 4 to 5 years in the autumn on the ground.

Fertilizers can also be supplied by the foliage, and in olive trees this characteristic can be effectively exploited in order to satisfy the needs of the plants in situations of particular demands (lacks of microelements), or as integration of soil fertilization in the different phenological phases. This technique is considered to be a valid support to increase the nutrient levels and the crop yield, reducing competition among metabolic sinks (shoots, inflorescences and fruits) and increasing the absorption of nutrients through the roots (Cimato et al., 1990, 1991, 1994, 1995; Toscano et al., 2000; Toscano, 2008). It provide nutrients quickly, uses low amounts of fertilizer, can be combined with pesticide applications, is well suited to rain-fed olive trees or when ground fertilizations would be useless due to a lack of soil humidity. The advantages of this technique are manifold: timely intervention, nutrients are given at the moment of greatest necessity, and is effective in a short time, and allows an integral use of the administered element. If only a foliage solution is applied, several applications take place (Fernández-Escobar, 1999; Ben Mimoun et al., 2004). Some results demonstrate, however, that foliar fertilization cannot entirely replace nutrition through the roots, even though it permits a reduction of the fertilizer required to be applied to the soil (Fontanazza, 1988; Toscano et al, 2000).

Many authors have studied the efficiency of olive foliar nourishment and for specific nutrients good results have been achieved using urea solution (Cimato et al, 1991; 1994).

Potassium is easily absorbed and distributed through leaf tissues (California Fertilizer Association, 1998) and foliar application is helpful to satisfy plant requirement having a high efficiency (Inglese et al., 2002). Phosphorous is given during summer fruit growth, in order to be readily absorbed and translocated to the fruits for quality purposes, therefore its application is better in the form of a foliar fertilizer.

In addition, recent studies have assessed the effectiveness of some commercial products on different olive cultivars, which behave differently. A helpful example of this efficiency is shown using NutriVant (NV) foliar fertilizers in addition to soil fertilization and showed that better results are obtained on the 'Carolea', than on 'Nocellara del Belice' cultivar. This difference is more clear in the "off year" orchards, during which an increase in the vegetative parameters and yield entity, in comparison to the control tests, was recorded. Conversely, in the "on year", the NV test had good results on both observed cultivars (Toscano et al., 2002b; Toscano, 2005; Toscano & Godino, 2010).

In irrigated orchards it is possible to supply nutrients to the plant by watering systems (fertigation) (Toscano et al., 2002a). The advantages of such practice consist in the easiness of application and in the efficiency of fertilizers, being able to reduce the needs of fertilizers by up to 30% in comparison to soil distribution. Fertigation implies a sensitive reduction of the management costs both in terms of purchase, transport and distribution of fertilizers, enhancing their efficacy in order to grant a better nutritional level to the trees, to maximize yield, oil production and profitability.

5. IRRIGATION

The efficient use of water resources in agriculture is extremely important in order to improve the economical and environmental sustainability of agricultural activity. Mediterranean regions of Italy are characterized by a high evaporative demand of the atmosphere, water scarcity and increasing negative consequences of climate change. In Italy the rainfall can vary annually from less than 400 to over 800 millimeters, and the lack of precipitation that is often manifested during the summer, involves the use of irrigation during dry periods to ensure the constant productivity of olive orchards.

In traditional olive cultivation areas, characterized by water scarcity, rainfall and underground water resources are the only supplies for the olive tree water requirements. Rainfed olive groves, therefore, are characterized by low plantation density which allows the exploitation of an adequate soil volume by the root system, minimizing competition for water among plants.

The olive, a sclerophyllous evergreen tree, is able to tolerate the low availability of water in the soil by means of morphological and physiological adaptations acquired in response to coping with drought stress (Connor & Fereres, 2005; Bacelar et al., 2007). Under semi-arid conditions olive trees were able to restrict water loss by modulating stomatal closure at different levels of soil moisture and evaporative demand and show a non-balanced allocation of dry matter among the different plant organs, resulting in a reduction of the vegetative growth and a significant decline in the productive performance (low yield and alternate bearing behaviour) in favour of development of the root system. Indeed, olive tree roots can extend and go deep into the soil to exploit a wider soil volume (Fernández et al., 1991; Dichio et al., 2002). Olive plants maintain a high rate of photosynthesis during long drought stress periods. The high efficiency of the olive is also due to its ability to continue to absorb carbon dioxide and to produce carbohydrates in water deficit conditions that determine the complete stomatal closure and threaten the survival of other species *(*Xiloyannis & Dichio, 2006*)*. A higher photosynthetic rate under drought is a decisive factor for better drought tolerance in olive cultivars (Bacelar et al., 2007). Generally, when water is not restricting growth, plants invest a considerable fraction of photoassimilates in the expansion of photosynthetic tissues, maximising light interception and, as a consequence, growth (Dale, 1988). The capacity to withstand severe and prolonged drought periods, however, is negatively associated with olive tree growth and productivity, owing to the decrease of assimilates under water deficit conditions. Reductions in photosynthetic performance under water stress have also been observed by several authors (Inglese et al., 1999; Patumi et al., 1999; Tognetti et al., 2005; Bacelar et al., 2006; Lavee et al., 2007; Ben Ahmed et al., 2009).

Much research shows the productivity benefits of irrigation. Irrigation is highly effective in increasing yield and yield components such as fruit size, fruit number and oil content, moreover, irrigation affects the pulp-to-pit ratio, phenology and time of fruit maturation (Agabbio, 1978; Goldhamer et al., 1994;Michelakis et al., 1994; Inglese et al., 1996; Gòmez-Rico et al., 2006, 2007; Dag et al., 2008; Servili et al., 2007).

A proper soil water availability enhances vegetative growth, such as shoot length, allowing the olive trees to produce a higher number of buds able to provide the opportune basis for the next year's production (Patumi et al., 2002; D'Andria et al., 2004; Gucci et al., 2007;Ben-Gal et al., 2008;). Stress levels and water requirements are highly dependent on fruit load and best irrigation management must account for biannual bearing effects. Although biennial bearing is basically genetically determined, the degree to which it occurs is greatly affected by environmental conditions, especially the weather and cultivation practices (Pandolfi et al., 2000). Alternate fruit bearing occurs under both extensive and intensive growing conditions (Pannelli et al., 1996; Lavee, 2006). With irrigation, olive production can increase up to five times that of olive groves in dry arid climates, in the Italian climate on average a double production must be expected (Bini et al., 1997). Obviously the scale of production will depend on soil conditions, average rainfall, evapo-transpiration and temperatures, cultivars, planting distances and other cultural practices (Nuzzo et al., 1997). Proper management of irrigation, especially during the summer drought, keeps leaves in activities promoting fruit growth and accumulation of reserves in the various plant organs (Xiloyannis & Palese, 2001), in any case, table olives cannot be cultivated without irrigation.

5.1. Olive Tree Water Needs

The unitary water consumption of the olive tree, namely the quantity of water that must be transpired in order to synthesize a gram of dry substance or commercial product, have been estimated to be 1 liter of water by 1 m^2 of leaf, daily transpired in August. Such indexes, with opportune calculations, can be useful to help establish watering volumes. Best indications are drawn by the compilation of water budgets that, from the comparison among the entity of the rains and the losses of damp from the soil by evapo-transpiration, allow the determining of the water deficit or excess in the different periods of the year.

The criteria to be adopted in watering planning must be based on respect of the water requirement of the crop, and on the knowledge of the critical phases of the vegetative cycle of the plants, over that of the quantity of available water for irrigation, for the evaluation of the economic convenience of the intervention. As for all the other production factors, the economic principle of marginal productivity is in force also for water.

Olive tree water requirements are variable and depend upon factors such as soil type, climate, planting density, age of trees, cultural management (e.g. fertilizing, pruning) and the method of irrigation.

In the olive tree there are nevertheless some critical periods during the annual cycle, during which the plant mostly needs water. The first one extends from bud differentiation up to flowering and therefore to the fruit set: in these phases a water deficit can create trouble in flower development with a smaller number of flowers for inflorescence, increasing ovary abortion, and a lower fruit set. Generally during this period in Italy there are no deficiencies in soil water. Subsequently, at the second phase of fruit growth, corresponding to the pit hardening period, olive trees are most resistant to water deficit (Goldhamer et al., 1994; Moriana et al., 2007), on the contrary the third phase, when olive oil is accumulated, the olive tree again seems to be sensitive to water stress (Lavee & Wodner, 1991)

The inolition process starts around the pit hardening phase and reaches a maximum before ripening. The effects of irrigation on oil content are nevertheless quite controversial depending on different experimental conditions

Some authors did not find any difference in oil content between irrigated and non-irrigated trees (Michelakis et al., 1994; d'Andria et al., 2004), while Inglese et al. (1999) reported a lower oil content in the fruits of trees grown under high soil water deficit conditions. The literature suggests that the fruit and oil yield response to irrigation is highly cultivar specific (Lavee et al., 2007). Despite the increasing use of irrigation in olive groves, there is still a poor understanding of the effect of irrigation deficit on the qualitative parameters of olive oil.

Increasing irrigation leads to fruits with a greater water content (lower oil percentage), and irrigation has been found to decrease the polyphenol content (Patumi et al., 1999; Gómez-Rico et al., 2006; Ben-Gal et al., 2008; Dag et al., 2008), which then changes the oil bitterness and spicy tastes.

Several studies, which focused on the effect of irrigation on olive oil composition, report that irrigation increases free fatty acids in oil (Dag et al., 2008), can affect the fatty acid composition (Ranalli et al.,1997; Aparicio & Luna 2002; Servili et al., 2007) and the accumulation of secondary metabolites, that are fundamental in improving the organoleptic characteristics of the oil, is increased (Pannelli et al., 1996; Inglese et al., 1996).

For the calculation of the water needs in an olive-grove, some formulas are used that consider climatic environmental data, such as the rains and the potential evapo-transpiration (ETP), adopting different coefficients in relation to the spacing of trees, the age and shape of the plants, and season. The water deficit, will be given by the difference between the water used by the crop and the water availability in the soil: such a deficit will be therefore compensated for with irrigation to optimize the productive potentialities of the plants.

The calculated seasonal watering volumes, will be nevertheless reduced considering the threshold of convenience, in relation to the efficiency of the irrigation system, the cost and availability of water and the value of the product. For the intensive olive growing in South Italy, it increased from 1.500 up to 3.000 m3 hectare^{-1} per year (Agabbio, 1978).

An evaluation of the water needs, such as the water consumed by the crop (evaporation and transpiration), can rationalize the irrigation technique. The evaporation potential (ET_0) must be determined through the compilation of the soil hydrological balance and the search for an empirical correlation between the potential evapo-transpiration and one of the climatic factors.

To satisfy the needs of an intensive olive-grove the results of different watering trials pointed out that for the olive tree it is enough to supply 30-50% of the evaporated water.

The beginning of the irrigation season should take place when the soil is still wet (60-70% of available water) to ensure the maintenance of adequate reserves even in deeper layers and at points not covered by providers in order, however, to maintain roots present in those areas.

Irrigation can be realized in different ways and the choice of the optimal method should be made according to each single olive-grove typology and environments. Sprinkling methods, with giant irrigators or wings, have the advantage of adapting to any soil condition, the facility of moving and transfer, and the timeliness of intervention, but generally with high costs and low efficiency of water. On the contrary localized irrigation, that allows water distribution evenly in sloping land, is a technique which offers the possibility to intervene in certain biologically critical phases for the plants (flowering, fruit setting, pit hardening, etc.), allowing a significant reduction, of about 25-30%, in the consumption of water. Furthermore it allows a more uniform distribution of water over time, with shorter shifts and increases the efficiency of irrigation up to 90%, avoiding losses due evapo-transpiration, runoff, etc.

With sprinkling the volumes are calculated for the whole surface; with the localized irrigation on the wet surface equal to 10% with drip irrigation and to 25% with microjets. Microjets enable irrigation of a rather large surface to meet the different needs of the olive tree during its development, but also creates constraints for tillage and weeds removal by mechanical means and increase water loss by evapo-transpiration.

In the center-northern olive-grove environments of Italy, natural water availability is often such to allow satisfactory production, even without resorting to irrigation. In the South of Italy, generally the annual average rainfall is rather low, with rains concentrated in the autumn-winter season, that does not coincide with the needs of the plant, therefore it is necessary to apply irrigation during summer.

The irrigation intervals depend, more than from the evaporative demand of the environment, on the type of soil and therefore from the quantity of water that it can retain.

In the case of localized irrigation shifts are on average 2-4 days with about 30 $m^3 ha^{-1}$ of water, the turns will progressively be shorter passing from slimy-sandy to sandy soil.

The most critical phases in which water stress should be avoided are at floraison, at fruit set, at fruit growth and at inolition. An effective watering season could start, according to watering water availability, at the end of flowering (May-June) and continuing until late September.

In recent years many studies have tried to apply deficit irrigation strategies to olive trees. These are based on the observation of Chalmers et al. (1981), who reported for peach trees that the maintenance of a slight plant water deficit can improve the partitioning of carbohydrate to reproductive structures, such as fruit, thus controlling excessive vegetative growth. The asynchronous growth of olive fruits and shoots reduces competition for assimilates at critical stages, providing a sound basis for the application of irrigation deficit.

The controlled irrigation deficit is a water management method that does not completely satisfy the tree's water requirements during the growing season. It causes a temporary and regulated water deficit in a specific phenological stage. When it is applied in the pit-hardening period, the olive oil yield is not affected while the water use efficiency (WUE) is improved. On the contrary when the controlled irrigation deficit is applied from fruit set to harvest, the oil yield decreases but the WUE and certain olive oil quality parameters improve. As the productive tree responses are not affected by moderate levels of water stress, irrigation deficit strategies are recommended in arid and semi-arid areas to save the scarce conventional water resources (Angelakis et al., 1999; Massoud et al., 2003).

Finally it is interesting to note that the olive is quite resistant to salinity. This plant tolerates brackish water (up to a salty residue of 4 g/liter), and therefore can allow the realization of irrigated olive-groves, valorizing waters which are not usable for other crops (Basta et al., 2002; Perica et al., 2008).

6. PRUNING

Pruning is a very expensive practice in olive grove management, reaching up to 40% of total cultivation costs, but it is also essential for olive grove profitability. It is finalized to modify the natural shape and structure of the trees, to reduce to the least one skeletal structures, to balance the vegetative and productive activity, and to maximize fructification. To reach the best results, pruning must be rationally managed, and based on the harvesting system. A modern approach to this practice allows to form and maintain the tree structure at a relatively low cost, reducing and simplifying pruning operations without negatively affecting yield, oil quality, or orchard sustainability.

Strategies of "minimum pruning" can be developed at a farm level independently of the type and size of the orchard. Managing the canopy according to the criteria of "minimum pruning" is suitable both for traditional olive groves and modern, high-density orchards.

The growing habit of the cultivar, the natural tendency for high vegetative activity, the type of buds and branches, and alternate bearing are all important biological features of the olive tree that it is important to consider for pruning.

In practice, pruning is distinguished into a formation and a production pruning: formation pruning has the purpose to give the selected form to the olive tree; while the production pruning is finalized to preserve the form and the size of the canopy, to eliminate inefficient or unproductive structures, to facilitate the functional positioning of fruiting shoots to enhance harvest efficiency, to maintain the trees' vegetative and productive balance. Olive trees bear fruits on the previous year's shoots, so to have fruit every year an adequate vegetative growth must be achieved. Annual interventions should be faster, smaller and easier cuts using small tools such as shears and saws.

The execution of pruning should avoid the accumulation of too much wood caused by an excess of primary branches and an excessive overlapping of secondary branches. To stimulate olive production pruning must be reduced to strictly necessary interventions, leaving the most possible greater number of leaves. Periodically some return cuts made on the branches return the plants to their assigned volume to maintain the volume and the shape of olive trees.

Extraordinary kinds of pruning are practiced when it is necessary to restructure the canopy in another form held to be more convenient. Pruning old trees requires drastic cuts to rejuvenate or to restore the health of the plants so as to stimulate their growth and renew fruit-bearing shoots and branches.

Pruning also contributes to reducing the occurrence of pest and disease. Dense canopies encourage the presence of parasites due to high relative humidity, whilst well-aerated canopies considerably decrease the attack of pest and disease such as the 'olive knot' which appears on branches and is otherwise very difficult to control.

6.1. Training System

Around the Mediterranean basin, a traditional area of olive tree cultivation, there are many different training systems for olive trees, but now in modern orchards the most common shapes are:

- The "vase" with several different variants is by far the most popular shape and the most practical for hand or semi-mechanical harvesting. Usually the vase has a single trunk varying in height from 50 cm to 120 cm, branching into some primary branches, in adult plants generally three, equally spaced so as to intercept as much sunlight as possible. These branches, tilted about 45-50 degrees, support the scaffolding of the tree.

- The only pruning required is in the centre of the canopy to allow enough sunlight to come through and removing cross branches leaving the greatest possible number of leaves on the plants, because productivity and the oil quantity in the drupes is dependent on them.

- The "single-trunk" is constituted by a central axis of the tree that raises a series of primary branches chosen amongst those that grow vigorously inserted in a spiral, alternated by 50-60 cm between them. To achieve this form few thinning cuts are made during the training phase. When the plants are in production the primary branches are periodically pruned by cutting or heads back, or eliminating them in order to renew the fruit-bearing surface. This shape is good for the intensive olive growing systems where the canopy cone reduces distances between the plants and for the mechanical harvest with shakers, but it is difficult to manage.

- The "bush" shape is the result of natural basal sprouting of the olive tree with numerous branches and those that arise from the bottom. This system is suitable for intensive cultivation models to be collected by hand or with tool facilitators it requires very little pruning during the training phase, but it is unsuitable for mechanical harvests with shakers due to lateral branches shooting from the proximal part of the trunk.

- The hedgerow is a training system in which trees grow freely, so that the canopy forms a productive wall along the row, usually managed with mechanical pruners to maintain the trees with an available volume.

Lately the training system suitable for olive orchards with over 1000 trees per hectare, is a single axis, obtained by thinning out the side branches in the apical part and by also removing those located below 0.5 m from soil during the first year of planting the to allow the passage of the machine. Once full production has been reached, plants are pruned so as to contain vegetation within 2.2 m in height and prevent the development of little branches of a diameter greater than 3 cm. In these groves mechanical pruning alternating with manual interventions is used to contain the development of foliage.

The training system is the result of the growing habit of the tree and pruning practices and it must be chosen before the planting as a function of the harvesting method and mainly of the area climate. Indeed, experimental tests of comparison between the "vase" and the "single-trunk" conducted over several years in the experimental field of CRA-OIL, located in Mirto Crosia (CS-Italy), showed the extreme difficulty in maintaining the last shape due to the climatic characteristics which are strongly favorable to olive plant development.

6.2. Pruning Scopes and Effects

A first objective of pruning is to provide a shape and structure to the tree which guarantees proper illumination of the canopy to enhance photosynthesis, good circulation of air, avoidance of pest disease, and a better disposition of fruiting shoots to facilitate and maximize the harvest.

At plantation, the first cuts are executed to plan the scaffold and the principal branches are chosen according to the selected shape. In the following years, pruning will be limited to the elimination of unfit shoots, favoring correct skeletal development. After the third-fourth year, with the beginning of the yield, annual pruning will have to balance volumetric growth with the vegetative-productive equilibrium of the trees.

Pruning intensity increases with the age of the tree. Pruning is light on young trees to allow the shaping and to grow and build energy reserves.

As general guidelines, for adult trees, given the light need of this species to perform photosynthesis, it is necessary to reduce the density of the foliage, allow sunlight to penetrate into every part of the olive canopy and promote air circulation. All the suckers around base of the trunk and branches that have already produced should be removed. It is also important to keep the upper parts of the tree open to allow the lower parts to remain productive.

In adult olive-groves, in full production, annual pruning allows better regulating of the balance between vegetative and reproductive activities and so contributes to reduce alternate bearing. This phenomenon is more dramatic the wider the pruning shift. The shift of pruning cycles should be chosen based on factors such as the growth of branches, the fruit-laden, the training system, soil fertility and climate and structural aspects of the farm. The execution of pruning every 2 or more years allows a reduction of the cost of such practice but favors the occurrence of alternate bearing.

Tests for several years showed that it is necessary to maintain a large leafy area of the plant, and pruning of medium intensity in shifts of two - three years, depending on the cultivar, gives the best results (Tombesi et al., 2002; Tombesi et al., 2007). These pruning cycles, compared with annual pruning, allow an increase in the production efficiency of the plant, and also a net saving of human work, since the execution times are not very different, and the number of cuts per plant is almost similar.

According to whether the mechanical harvest are adopted, with shakers or mechanization, in the first instance it is necessary to build the canopy towards erect forms, with shoots which are relatively short and rigid, to favor the transmission of vibrations up to the drupes in the whole volume of the canopy, while in the second case, the pruning will have the aim of bringing up the external wall of the canopy longer and pending fruiting shoots, to facilitate harvesting with pneumatic combs(De Simone &Tombesi, 2006; Tombesi et al., 2008).

Pruning must be carried out at the end of winter, before the restart of vegetative growth. It must be avoided after the harvest, because it reduces the cold resistance of the plants and does not allow wounds to heal, favoring diseases from fungi or other parasites. Traditionally in late summer a second pruning is performed on adult plants to eliminate suckers inside the plants, with special reference to the vase shape, where the formation of shoots within the canopy becomes a rule.

In profitable olive orchards pruning mechanization is essential to reduce management costs and regain timeliness in working, even though it penalizes the productive efficiency of trees. To balance the economic needs of management there are several ways to carry out mechanical pruning that should not be performed simultaneously on the entire plant. To ensure a good production of fruiting branches the canopy that remains after pruning should not excessively reduced. Mechanical pruning can be applied alternately in different years and/or rows by carrying out topping and hedging, reserving annual hand pruning to eliminate shoots and suckers.

Several experiments have been performed in different types of olive orchards in order to study the feasibility of mechanical pruning (Giametta & Zimbalatti, 1997; Ferguson et al., 1999; Ferguson et al., 2002; Peça et al., 2002; Tombesi et al, 2008; Dias et al., 2008; Farinelli et al., 2009). In our experience, the technical-economical convenience of pruning mechanization, also considering equipment integrated with pneumatic combs for olive harvesting, was evaluated. The results showed the good effectiveness of the pruning equipment in terms of cutting quality and working capacity (Pascuzzi et al., 2007; Toscano 2010a). Other trials have been carried out to assess the performances and the effects of mechanical pruning, that was performed both with toothed disks and scissor blades pruning machines. Both resulted in great efficacy and work productivity; nevertheless, the first ones are more efficient for woody vegetation, up to 10 cm diameter, while the second ones work better on thin branches, even though it can cut woody branches up to 5 cm diameter (unpublished data).

7. HARVESTING

In olive grove management, the harvest is the other most expensive practice, together with pruning. Harvesting systems, can be considered rational only when they can reconcile the operation costs with the necessity to pick up the maximum yield in respect of the product quality. In the past, the availability of low cost manpower with a manual harvest allowed the satisfaction of these two demands, but the low availability and increase of the cost of labor, have made such operations excessively onerous, and applicable only to table olives.

Manual harvest can be improved using hand-held pneumatic combs to detach the olives from the plant that assuage the work, and give maximum flexibility in terms of harvesting time, and increase the operators productivity, but it is time-consuming and costly.

Mechanical harvesting is executed with shaker, also equipped with a reverse umbrella as an olive interceptor, that has considerable economic advantages compared with traditional manual picking procedures (Tombesi & Tombesi, 2007). In this way a great reduction in labour costs, harvesting timeliness and good performance, is achieved. Nevertheless it is difficult to apply in the majority of traditional olive groves due to the presence of malformed, voluminous plants, or those of an unsuitable cultivar.

In the new intensive olive groves, with trees optimized for cultivar and structure, mechanical harvest is instead applicable with positive results, usually in a step, getting up to 80-90% of yield (Hartmann & Reed, 1975; Ferguson et al., 1999; Giametta & Pipitone, 2004; Toscano & Casacchia, 2006). In these orchards with well pruned trees it is possible to harvest up to 50 trees/hour with a suitable shaker and collecting system (Lavee, 2010).

Using shakers, the more efficient harvesting yard is constituted by 5 or 7 operators, of which one operates the shaker, and the others the nets and the moving of olives, reaching a productivity up to 0,4 tons/man by hour (Briccoli et al., 2006; Tombesi, 2006; Toscano, 2007).

In super intensive olive orchard, or intensive olive orchard with trees structured in a productive wall, continuous harvesters, derived from grape pickers that work on both walls of a row (Bellomo et al., 2003; Arrivo et al., 2006) can be used. Wall pickers, that work on a single side of tree walls, also reach a working productivity up to 1 hectare by hour, with yield percentages similar to the shakers (Toscano, 2010b).

7.1. Ripening Physiology of Drupes

The olive tree fruits (Olea europaea L.) are oval or round drupes, of variable weight according to the cultivar, the yield, the nutritional and health state of the tree. The drupe is constituted by the external exocarp (peel), by the fleshy mesocarp (pulp) and by the internal stone (pit). Fruit development and ripening are a combination of biochemical and physiological changes that occurred during maturity of fruit. The development process is characterized by changes in size, weight, composition, color, flavor and physical proprieties of the fruit (Connor & Fereres, 2005) and is critical for final yield and oil quality. Oil accumulation, which occurs from pit hardening to harvesting may be early or late depending on the variety, generally it starts in the northern hemisphere from the month of August and continues up to November-December, subsequently the increase of oil content in the drupes is apparent being due to water reduction.

The maturation of olives also varies depending on the crop load, environmental conditions, which are subject to strong annual variations, soil moisture, and cultural practices. At harvest, within a tree, not all the fruit are at the same ripening stage, in fact this parameter also depends on the position of the fruits.

For the olives to be used for the oil extraction, the harvest must have been carried out at the beginning of the ripening phase of fruits, at veraison, when the pulp changes its color from green to purplish red. This stadium corresponds to the maximum oil yield per hectare, since, even if subsequently a slight increase in the oil content of the drupes is had, the loss for natural fruit drop undoes the advantage. In many instances the oil quality also decreases. An early harvest allows the production a good oil, rich in antioxidants and aromatic flavors, that confers resistance to oxidation, and a "fruited" taste. Instead, oils obtained from olives harvested at an advanced ripening stage are less intense, less bitter with a lower percentage of mono-unsaturated and saturated fatty acids and a higher percentage of poly-unsaturated fatty acids, that penalize its stability.

With regards to the harvesting method, oils with excellent quality can be produced with both manual and mechanical harvesting, as long as the drupes are intact and healthy.

For table olives harvesting is carried out manually from the plant, to avoid damage to the fruits that would consequently depreciate their market value. In some cases harvesting in the olive-grove is done in different steps as a consequence of the ripening scale, or performed at the same time and the sub-size fruits are sent to the crusher for oil extraction. To facilitate the harvest it is necessary that the plants are of a contained dimension, and with suitable forms that assemble the fruits on the outside of the canopy and on lean shoots.

In table cultivars for green fruit processing the harvest is done when the peel color changes to light green, that corresponds to the beginning of pulp softening and the maximum content of sugars, fundamental during tanning, for the fermentation process that follows sweetening, with lye or brine. For the olives destined for tanning to black, the harvest must be effected when the pulp is also colored, based on the physiological maturation of the fruits.

The most important parameters to determine the stage of maturation of the drupes are the fruit separation force and the development of natural fruit drop. Before the natural fall of olives there is an attenuation of the force with which they are attached to branches and shoots. These physiological changes do not occur simultaneously on all the drupes of the same tree but occur with a certain scaling. Therefore, the decrease of the attachment strength of the fruit and the drop of the first fruits are the most important indices of the final stage of ripening. These indices are easily determined and able to predict with sufficient reliability the time to start harvesting.

The optimal time of harvest can be, further, defined as one in which there is a high amount of fruit on the plant capable of being detached by the machines in considerable percentages and with a high content in good quality oil when ripening the pulp becomes less consistent (Farinelli et al., 2006).

8. CONCLUSIONS

The information contained in this paper highlights that it is possible to achieve some improvements in olive tree productivity (in terms of quantity and quality) and a reduction in costs, spreading more rational agronomic practices. Increasing the olive groves income can be achieved through updated cultivation techniques. All these must be coordinated and integrated with each other to obtain a rapid formation of the tree production structures that allow the maximum expression of their productive potential and provide a high level of mechanization. Soil management, plant structure, fertilizer, irrigation, pruning and mechanical harvesting must be chosen according to variety and environmental features. The paper provides useful indications on the introducing of the cover crop to better soil management in order to control erosion and maintain soil fertility.

Moreover, olive trees respond very strongly to irrigation and take advantage of very low volume of water also with regulated irrigation deficit. With regards to olive pruning, this cultural practice must be managed rationally based on the harvesting system and both these techniques (pruning and harvesting) must be done mechanically to reduce the running costs for better crop competitiveness.

ACKNOWLEDGEMENT

Financial support for this study was provided by the Italian Ministry of Agriculture, Food and Forestry Policy through the project GERMOLI "Salvaguardia e valorizzazione del GERMoplasma OLIvicolo delle collezioni del CRA-OLI".

REFERENCES

1. AA.VV.1989Paolo Sequi (a cura di), Chimica del suolo,. Pàtron (ed.),Bologna, Italy.

2. M. Agabbio, 1978Influenza dell'intervento irriguo sul ciclo produttivo dell'olivo. Nota II: Influenza del regime idrico sulla biologia fiorale e sui caratteri morfo- qualitativi dei frutti. Studi Sassaresi. Sez. III. Vol. XXV: 266272

3. N. Angelakis, Monte. M. H. F. Marecos, L. Bontoux, T. Asano, 1999The status of wastewater reuse practice in the Mediterranean basin: need for guidelines. Water Res. 1022012217

4. R. Aparicio, G. Luna, 2002Characterisation of monovarietal virgin olive oils. Eur. J. Lipid Sci. Tech. 104614627

5. Arrivo, F. Bellomo, P. D'Antonio, 2006Raccolta meccanica dell'oliveto superintensivo. Inf. Agrario, 1, 6871

6. E. A. Bacelar, D. L. Santos, J. M. Moutinho-Pereira, B. C. Gonçalves, H. F. Ferreira, C. M. Correia, 2006Immediate responses and adaptative strategies of three olive cultivars under contrasting water availability regimes: changes on structure and chemical composition of foliage and oxidative damage. Plant Sci 170596605

7. E. A. Bacelar, J. M. Moutinho-Pereira, B. C. Gonçalves, H. F. Ferreira, C. M. Correia, 2007Changes in growth, gas exchange, xylem hydraulic properties and water use efficiency of three olive cultivars under contrasting water availability regimes. Environ. Exp. Bot. 60183192

8. P. Basta, P. Toscano, A. Ciliberti, D. Turco, 2002Osservazioni pluriennali sulle risposte vegeto-produttive di cinque cultivar di olivo irrigate con acqua salmastra a differenti concentrazioni. Conv. Internaz. di Olivicoltura, Centenario I.S.Ol. Spoleto, (PG) 235240

9. Belsito, et al.1988Chimica agraria. Zanichelli (Ed.) Bologna, Italy.

10. F. Bellomo, P. D'Antonio, F. D'Emilio, 2003Spagna, quando l'olivicoltura è superintensiva. Olivo e Olio, 11813

11. Ahmed. C. Ben, Rouinab. B. Ben, S. Sensoyc, M. Boukhrisa, Abdallah. F. Ben, 2009Changes in gas exchange, proline accumulation and antioxidative enzyme activities in three olive cultivars under contrasting water availability regimes. Environ. Exp. Bot. 67345352

12. Ben-Gal, A. Dag, U. Yermiyahu, Y. Zipori, E. Presnov, I. Faingold, Z. Kerem, 2008Evaluation of irrigation in a converted, rain fed olive orchard: the transition year. Acta Hortic. 79299106

13. Mimoun. M. Ben, O. Loumi, M. Ghrab, K. Latiri, R. Hellali, 2004Foliar Potassium Application On Olive Tree M. IPI regional workshop on Potassium and Fertigation development in West Asia and North Africa; Rabat, Morocco, 2428November, 2004.

14. G. Bini, L. Sapia, Bati. C. Briccoli, P. Toscano, 1997Irrigazione dell'olivo con sistemi a microportata: risultati sulla cv Carolea in Calabria. Inf. Agrario, LIII, 223744

15. H. Blank, 1932Handbuch der Bodenlehre. 7 volumes, Springer Verlag, Berlin.

16. F. Bonciarelli, 1981Agronomia. Edagricole (Ed.), Bologna, Italy.

17. Bouat, 1968Physiologie de l' olivier et analyse des feuilles. Inf. Oleic. Int. 427393

18. Bati. C. Briccoli, R. Rinaldi, T. Sirianni, 1992Prime osservazioni sulla presenza di micorrize di tipo VAM in oliveti dell'Italia meridionale. In Atti Giornate Scientifiche SOI- 1992. Ravello, Italy, 4647

19. Bati. C. Briccoli, C. Tocci, P. Toscano, F. Gazea, 1995Influenza di potatura ed irrigazione sullo stato nutrizionale dell'olivo. Conv. "Tecniche, Norme e Qualità in olivicoltura". Potenza, Italy, 305316

20. Bati. C. Briccoli, P. Toscano, S. Antonuccio, O. Failla, 2002Effetto dell'inerbimento sullo stato vegeto-produttivo e nutrizionale dell'olivo. Inf. Agrario, anno LVIII. 31

21. Bati. C. Briccoli, P. Toscano, M. L. Perrotta, 2006La raccolta meccanica in diverse tipologie olivicole calabresi. Conv. Naz. "Maturazione e raccolta delle olive: strategie e tecnologie per aumentare la competitività in olivicoltura". Alanno (PE), Italy. 9599

22. California Fertilizer Association,1998Western Fertilizer Handbook, Second horticultural edition, Sacramento, CA. 362

23. R. Calvente, C. Cano, N. Ferrol, C. Azcón-Aguilar, J. M. Barea, 2004Analysing natural diversity of arbuscular mycorrhizal fungi in olive tree (Olea europaea L.) plantations and assessment of the effectiveness of native fungal isolates as inoculants for commercial cultivars of olive plantlets. Applied Soil Ecology, 261119

24. D. J. Chalmers, P. D. Mitchell, L. Vanheek, 1981Control of peach tree growth and productivity by regulated water supply, tree density and summer pruning. J. Am. Soc. Hortic. Sci. 106307312

25. Cimato, M. Marranci, M. Tattini, 1990The use of foliar fertilization to modify sinks competition and to increase yield in olive (Olea europaea cv Frantoio). Acta Hort. 286175178

26. Cimato, M. Marranci, M. Tattini, G. Sani, 1991La concimazione fogliare come mezzo per incrementare la produzione nell'olivo. Inf. Agrar. 147173

27. Cimato, G. Sani, L. Marzi, M. Marranci, 1994Efficienza e qualità della produzione in olivo: riflessi della concimazione fogliare con urea. Olivae n. 544855

28. Cimato, M. Marranci, L. Marzi, G. Sani, 1995Riflessi della concimazione fogliare sulla produttività dell'olivo. Atti conv. su: "Tecniche, norme e qualità in olivicoltura". Potenza (Italy). 137148

29. D. J. Connor, E. Fereres, 2005The physiology of adaptation and yield expression in olive. Hortic. Rev. 31155229

30. R. d'Andria, A. Lavini, G. Morelli, M. Patumi, S. Terenziani, D. Calandrelli, F. Fragnito, 2004Effects of water regimes on five pickling and double aptitude olive cultivars (Olea europaea L.). J. Hortic. Sci. Biotech. 791825

31. Dag, A. Ben-Gal, U. Yermiyahu, L. Basheer, N. Yogev, Z. Kerem, 2008The effect of irrigation level and harvest mechanization on virgin olive oil quality in a traditional rain-fed "Souri" olive orchard converted to irrigation. J. Sci. Food Agric. 8815241528

32. J. E. Dale, 1988The control of leaf expansion. Annual Rev. Plant Physiol. 39267295

33. G. De Simone, A. Tombesi, 2006Influenza delle caratteristiche strutturali della pianta di olivo sulle rese di raccolta con l'uso di vibratori di tronco. Atti Conv. Naz. "Maturazione e raccolta delle olive: strategie e tecnologie per aumenrtare la competitività in olivicoltura". Alanno (PE), Italy 6368

34. R. Dexter, 1988Advances in characterization of soil structure. Soil Tillage Research 11199239

35. B. Dias, J. Peça, A. Pinheiro, L. Santos, N. Morais, A. G. Pereira, 2008The influence of mechanical pruning on olive production and shaker efficiency. Acta Horticulturae 791307313

36. Dichio, M. Romano, V. Nuzzo, C. Xiloyannis, 2002Soil water availability and relationship between canopy and roots in young olive trees (cv. Coratina). Acta Horticulturae, 586255258

37. European Soil Bureau,1999Database georeferenziato dei suoli europei. Manuale delle procedure. ESB, Comitato Scientifico. Versione 1.1. Versione italiana. E. A.C. Costantini (ed.). JRC, Ispra (VA), 170

38. O. Failla, A. Scienza, G. Stringari, D. Porro, S. Tarducci, N. Bazzanti, M. Toma, 1997Diagnostica fogliare per l'olivicoltura toscana. Informatore agrario, n. 396371

39. Farinelli, M. Boco, M. Ruffolo, A. Tombesi, 2006Ottimizzazione del periodo di raccolta per aspetti agronomici, meccanici e di qualità dell'olio. Atti Conv. Naz. "Maturazione e raccolta delle olive: strategie e tecnologie per aumentare la competitività in olivicoltura". Alanno (PE), Italy, 175182

40. Farinelli, M. Ruffolo, S. Tombesi, 2009Influenza della potatura meccanica sulla produzione e sulla qualità dell'olio di olivi allevati a vaso. Rivista di Frutticoltura e di Ortofloricoltura 117681

41. L. Ferguson, R. Heraclio, P. Metheney, 1999Mechanical harvesting and hedging of California black ripe (Olea europea L.) cv. 'Manzanillo' table olive. Acta Hort. 474193196

42. L. Ferguson, W. H. Krueger, H. Reyes, P. Metheney, 2002Effect of mechanical pruning on (Olea europea L.) California black ripe cv. 'Manzanillo' table olive. Acta Hort., 586, 281284

43. J. E. Fernández, F. Moreno, F. Cabrera, J. L. Arrue, J. Martin-Aranda, 1991Drip irrigation, soil characteristics and the root distribution and root activity of olive trees. Plant Soil 133239251

44. R. Fernández-Escobar, 1999Fertilización. In: Barranco, D., Fernández-Escobar, R., Rallo, L. (Eds.), El cultivo del olivo. Mundi-Prensa, Madrid, 247265

45. G. Fontanazza, 1988Olivicoltura la ripresa si fonda su concimazione ed irrigazione adeguate. Giornale di Agric. 153639

46. M. Freeman, K. Uriu, H. T. Hartmann, 1994Diagnosing and correcting nutrient problems. In. Ferguson, L., Sibbett, G.S., Martin, G.C. (Eds.) Olive production manual. University of California, USA, pubb. 33537786

47. G. Giametta, G. Zimbalatti, 1997Mechanical pruning in new olive-groves. J. Agr. Eng.Res. 681520

48. G. Giametta, F. Pipitone, 2004La raccolta delle olive. Innovazioni tecnologiche. Conv. UNACOMA e Accademia Nazionale di Agricoltura: Innovazione tecnologica e ricerca per lo sviluppo di una moderna meccanizzazione agricola, Bologna, Italy, 7993

49. L. Giardini, 1986Agronomia generale. Pàtron (Ed.), Bologna, Italy.

50. Giordano, 1999Pedologia. UTET, Torino, Italy

51. G. Gisotti, 1988Principi di geopedologia. Calderini, Bologna, Italy.

52. Goldhamer, J. Dunai, L. Ferguson, S. Lavee, I. Klein, 1994Irrigation requirements of olives trees and responces to sustained deficit irrigation. Acta Hort. 356172175

53. Gómez-Rico, M. D. Salvador, M. La Greca, G. Fregapane, 2006Phenolic and volatile compounds of extra virgin olive oil (Olea europaea L. Cv. Cornicabra) with regards to fruit ripening and irrigation management. Agric. Food Chem. 5471307136

54. Gómez-Rico, M. D. Salvador, A. Moriana, D. Pérez, N. Olmedilla, F. Ribas, G. Fregapane, 2007Influence of Different Irrigation Strategies in a Traditional Cornicabra cv. Olive Orchard on Virgin Olive Oil Composition and Quality. Food Chem. 100568578

55. R. Gucci, E. M. Lodolini, H. F. Rapoport, 2007Productivity of olive trees with different water status and crop load. J. of Hort. Sci. and Biotechnology, 824648656

56. Hakansson, W. B. Voorhees, H. Riley, 1988Weather and other environmental factors influencing crop responses to tillage and traffic. Soil Tillage Research 11239282

57. X. Hao, B. C. Ball, J. L. B. Culley, M. R. Carter, G. W. Parkin, 2008Soil density and porosity. In: Carter, M.R. & Gregorich, E.G. (Eds.) Soil sampling and methods of analysis. CRC Press.

58. H. T. Hartmann, 1958Some responses of the olive to nitrogen fertilzers. Proc. Am. Soc. Hort. Sci. 72257266

59. H. T. Hartmann, W. Reed, 1975Mechanical harvesting of olives. California Agriculture, Jou., 7546

60. W. H. Hendershot, H. Lalande, M. Duquette, 2008aIon exchange and exchangeable cations. In: Carter M.R. and Gregorich E.G. (Eds.) Soil sampling and methods of analysis. CRC Press.

61. W. H. Hendershot, H. Lalande, M. Duquette, 2008bSoil reaction and exchangeable acidity. In: Carter M.R. and Gregorich E.G. (Eds.) Soil sampling and methods of analysis. CRC Press.

62. P. Inglese, E. Barone, G. Gullo, 1996The effect of complementary irrigation on fruit growth, ripening pattern and oil characteristic of olive (Olea europaea L.) cv. Carolea. J. Hort. Sci. 712257263

63. P. Inglese, G. Gullo, L. S. Pace, 1999Summer drought effects on fruit growth, ripening and accumulation and composition of 'Carolea' olive oil. Acta Hortic. 474269273

64. P. Inglese, G. Gullo, L. S. Pace, 2002Fruit growth and olive quality in relation to foliar nutrition and time of application. Acta Hort. 586507509

65. P. Jeffries, J. M. Barea, 2001Arbuscular mycorrhiza-a key component of sustainable plant-soil ecosystems. In: Hock, B. (Ed.) The Mycota IX Fungal Associations. Springer, Berlin, 95113

66. J. B. Jones, 2003Agronomic handbook. CRC Press.

67. Kabata-Pendias, A. B. Mukherjee, 2007Trace elements from soil to human. Springer-Verlag, Berlin Heidelberg.

68. Kay, 1990Rates of change of soil structure under different cropping systems. Adv. Soil Sci. 12141

69. S. Lavee, M. Woodner, 1991Factor affecting the nature of oil accumulation in fruit of olive (Olea europaea L.) cultivars. J. of Hort. Sci. 66583591

70. S. Lavee, 2006Biennial bearing in olive (Olea europaea L.). In FAO (Ed.) Olea. FAOOlive Netw 25513

71. S. Lavee, E. Hanoch, M. Wodner, H. Abramowitch, 2007The effect of predetermined water deficit on the performance of cv. Muhasan olives (Olea europea L.) in the eastern coastal plain of Israel. Scientia Hort. 112156163

72. S. Lavee, 2010Integrated mechanical, chemical and horticultural methodologies for harvesting of oil olives and the potential interaction with different growing systems. A general review. Avd. Hort. Sci. 241515

73. P. E. Lovato, V. Gianinazzi-Pearson, A. Trouvelot, S. Gianinazzi, 1996The state of art of mycorrhizas and micropropagation. Adv. Hortic. Sci. 104652

74. M. A. Massoud, M. D. Scrimshaw, J. N. Lester, 2003Qualitative assessment of the effectiveness of the Mediterranean action plan: wastewater management in the Mediterranean region. Ocean Coast. Manage. 46875899

75. N. I. C. Michelakis, E. Vouyoucalou, G. Clapaki, 1994Plant growth and yield response of the olive tree cv Kalamon, to different levels of soil water potential and methods of irrigation. Acta Hortic. 356205210

76. J. J. Miller, D. Curtin, 2008Electrical conductivity and soluble ions. In Carter, M.R. & Gregorich, E.G. (Eds), 2008. Soil sampling and methods of analysis. CRC Press.

77. Moriana, D. Pérez-López, A. Gómez-Rico, Desamparados. de los, M. Salvador, N. Olmedilla, Ribas, Fregapane, G. (2007Irrigation scheduling for traditional, low-density olive orchards: water relations and influence on oil characteristics. Agric. Water Manage. 87171179

78. S. Natali, 1993Cure colturali dell'oliveto. Inform. Agrar. n.236771

79. V. Nuzzo, C. Xiloyannis, B. Dichio, G. Montanaro, G. Celano, 1997Growth and yield in irrigated and non-irrigated olive trees cultivar Coratina over four years after planting. Acta Hortic., 449, 7582

80. M. Palese, G. Celano, B. Dichio, V. Nuzzo, C. Xiloyannis, 1997Esigenze nutrizionali dell'olivo in fase di allevamento. Informatore Agrario, 44, 3540

81. Pandolfi, P. Proietti, F. Famiani, P. Guelfi, D. Farinelli, G. Sisani, A. Tombesi, 2000Effetti del tipo di gestione del suolo e dell'irrigazione sull'attività vegeto produttiva dell'olivo in centro Italia. V giornate scientifiche SOI 2000, Sirmione, Italy 329330

82. Pannelli, R. Selvaggini, M. Servili, M. Baldioli, G. F. . Montedoro, 1996La produzione e la composizione dell'olio in relazione alla fisiologia dello stress idrico in olivo (Olea europaea L.) cv Leccino. In: Atti convegno 'Olivicoltura mediterranea: stato e prospettive della coltura e della ricerca'. Cosenza, Italy, 701721

83. S. Pascuzzi, P. Toscano, A. Guarella, 2007La meccanizzazione della potatura di riforma degli oliveti- Atti Conv. Naz.- Tecnologie innovative nelle filiere:orticola, vitivinicola, e olivicola-olearia- Pisa-Volterra, Italy, Vol. III: 7780

84. M. Pastor, 1989Incidenza delle male erbe sull'evoluzione del contenuto d'acqua del suolo in oliveto di seccagno. Olivae 283237

85. M. Patumi, R. D'Andria, G. Fontanazza, J. Morelli, P. Giorgo, G. Sorrentino, 1999Yield and oil quality on intensively trained trees of three cultivars of olive (Olea europea L.) under different irrigation regimes. J. Hortic Sci Biotechnol. 74729737

86. M. Patumi, R. D'Andria, V. Marsilo, G. Fontanazza, G. Morelli, B. Lanza, 2002Olive and olive oil quality after intensive monocone olive growing (Olea europaea L., cv. Kalamata) in different irrigation regimes. Food Chem., 772734

87. J. O. Peça, A. B. Dias, A. C. Pinheiro, L. Santos, N. Morais, A. G. Pereira, D. R. De Souza, 2002Mechanical pruning of olive trees as an alternative to manual pruning. Acta Hort. 586295299

88. S. Perica, N. Bellaloui, C. Greve, H. Hu, P. H. Brown, 2001Boron Transport and Soluble Carbohydrate Concentrations in Olive. J. Amer. Soc. Hort. Sci. 1263291296

89. S. Perica, S. Goreta, G. V. Selak, 2008Growth, biomass al location and leaf ion concentration of seven olive (Olea europaea L.) cultivars under increased salinity. Scientia Hort., 117, 124129

90. Petruccioli, M. V. Parlati, 1983La concimazione dell'olivo. Terra e vita. anno XXIV, 35559

91. M. Raglione, P. Toscano, R. Angelini, Bati. C. Briccoli, M. Spadoni, C. De Simone, P. Lorenzoni, 2000Olive yield and soil loss in hilly environment of Calabia (Southern Italy). Influence of permanent cover crop and ploughing. VI International meeting on soils with mediterranean type of climate (IMSMTC). Barcellona (Catalonia), 4-9 Jul 1999, 10381040

92. Ranalli, G. De Mattia, M. L. Ferrante, L. Giansante, 1997Incidence of olive cultivation area on the analytical characteristics of the oil. Note 1. Riv. Ital. Sost. Grasse 74501508

93. V. Rasiah, B. D. Kay, 1995Runoff and soil loss as influenced by selected stability parameters and cropping and tillage practices. Geoderma, 68321329

94. E. Roldán-Fajardo, J. M. Barea, 1985Mycorrhizal dependency in the olive tree (Olea europaea L.). In: Gianinazzi-Pearson, V., Gianinazzi, S. (Eds.) Physiological and Genetical Aspects of Mycorrhizal. INRA, Paris, 323326

95. M. Servili, S. Esposto, E. Lodolini, R. Selvaggini, A. Taticchi, S. Urbani, G. Montedoro, M. Serravalle, R. Gucci, 2007Irrigation effects on quality, phenolic composition and selected volatiles of virgin olive oils cv Leccino. J. Agric. Food Chem. 5566096618

96. S. Soyergin, I. . Moltay, Ç. Genç, A. E. Fidan, A. R. Sütçü, 2000Nutrient status of olives grown in the Marmara region. Acta Horticulturae 586375380

97. O. R. Stein, W. H. Neibling, T. J. Logan, W. C. Moldenhauer, 1986Runoff and soil loss as influenced by tillage and residue cover. Soil Sci. Soc. Am. J. 5015271531

98. R. Tognetti, R. D'Andria, G. Morelli, A. Alvino, 2005The effect of deficit irrigation on seasonal variations of plant water use in Olea europaea L. Plant Soil 273139155

99. Tombesi, N. Michelakis, M. Pastor, 1996Raccomandazioni del gruppo di lavoro di tecniche di produzione in olivicoltura e produttività. Olivae, 633851

100. Tombesi, M. Pilli, M. Boco, 2002Prime valutazioni su intensità e periodicità della potatura nell'olivo. Rivista di Frutticoltura 107176

101. Tombesi, 2006Planting systems, canopy management and mechanical harvesting. OliveBioTecq, Recent advances in olive industry. Mazara del Vallo, Italy: 307316

102. Tombesi, P. Guarella, C. Di Vaio, P. Toscano, 2007Innovazioni nella meccanizzazione della raccolta e della potatura e ristrutturazione degli impianti olivicoli. Atti Convegno internazionale Ricerca ed innovazione per la filiera olivicolo-olearia dei Paesi del Mediterraneo. AGRILEVANTE, Bari. Pgt RIOM, Tomo I, Olivicoltura, Guido Ed. Rende,175190

103. Tombesi, D. Farinelli, M. Ruffolo, A. Sforna, 2008Primi risultati sulla meccanizzazione agevolata ed integrale della potatura dell'olivo. In: Atti Con. COM.SI.OL "Competività del Sistema Olivo in Italia", Spoleto, Italy 141147

104. Tombesi, S. Tombesi, 2007Olive harvesting and mechanization. In International Olive Council (Ed.) "Production Techniques in Olive Growing". Madrid, Spain 317346

105. P. Toscano, 1998MANOLI: Ipermanuale di olivicoltura. Progetto IIRA (Ipertesti, Internet e Ricerca Agraria), INEA(Ed.), Roma, Italy.

106. P. Toscano, Bati. C. Briccoli, T. Trombino, 1999Grass-cover effects on the vegetative and productive state of a young hilly olive-grove. Acta Horticolturae, 474181184

107. P. Toscano, 2000Gestione del suolo: Ciclo delle lavorazioni, inerbimenti e controllo della flora spontanea degli oliveti. Conv. Rassegna olio qualità ambiente. Metodi e sistemi innovativi dell'olivicoltura biologica e sostenibile. Stato della ricerca e della sperimentazione. ISO Rende, Italy 7383

108. P. Toscano, Bati. C. Briccoli, R. Sirianni, 2000Tecnica colturale: effetti delle concimazioni fogliari in olivicoltura. V Giornate Scientifiche S.O.I., Vol II, Sirmione (BS) Italy 583584

109. P. Toscano, A. Madeo, G. Godino, 2002aRisposte di tre cultivar di olivo a differenti tipi di concimazione in un ambiente litoraneo del Sud Italia. Conv. Internaz. di Olivicoltura. Centenario I.S.Ol. Spoleto (PG), Italy, 310315

110. P. Toscano, G. Godino, T. Belfiore, Bati. C. Briccoli, 2002bFoliar fertilization: a valid alternative for olive cultivar. Acta Horticolturae, 594, 191195

111. P. Toscano, Bati. C. Briccoli, G. Godino, C. De Simone, M. Raglione, P. Lorenzoni, R. Angelini, S. Antonuccio, 2004aEffetti Agronomici e pedologici di due diverse tecniche di gestione del suolo in un oliveto collinare del Meridione d'Italia. Olivae 1022126

112. P. Toscano, B. Scazziota, C. Briccoli-Bati, 2004bStrategie di gestione eco-compatibile per il recupero di competitività in sistemi olivicoli marginali: esperienze sull'inerbimento in suolo declive ed effetti sugli aspetti vegeto-produttivi su cv Carolea. In atti Conv. "Il futuro dei sistemi olivicoli in aree marginali: aspetti socio-economici, gestione delle risorse naturali e produzioni di qualità". Matera, Italy, 399407

113. P. Toscano, 2005Foliar fertilization on olive growing: first results of a specifics foliar fertilizers applied on some cultivars in different environments of Southern Italy. FAO Land And Water Bulletin, 10, 127130

114. P. Toscano, T. Casacchia, 2006La raccolta meccanica dell'olivo: effetti dell'Olive HarvestVant su Leccino in Calabria. In I.T.A.S. Cuppari (Ed.). Atti Conv. Naz. Maturazione e raccolta delle olive: strategie e tecnologie per aumentare la competitività in olivicoltura.-- Alanno (PE), Italy,101105

115. P. Toscano, T. Casacchia, Bati. C. Briccoli, B. Scazziota, 2006Inerbimento temporaneo e permanente in olivicoltura. Inf. Agrario, 21, 3639

116. P. Toscano, 2007Come organizzare i cantieri per l'olivicoltura calabrese. Olivo&Olio 11-12/076064

117. P. Toscano, 2008Concimazione fogliare insieme a quella al terreno. Olivo&Olio 35053

118. P. Toscano, 2009The controlled grass-cover soil management in southern Italy olive orchard environments. 18th Symposium Of The International Scientific Centre Of Fertilizers- More Sustainability In Agriculture: New Fertilizers And Fertilization Management. Rome, Italy, 381386

119. P. Toscano, 2010aPotatura meccanica solo ad anni alterni. Olivo&Olio, 34448

120. P. Toscano, 2010bRaccolta delle olive efficiente con macchine e cantieri adeguati. Inf. Agrario, 38284

121. P. Toscano, G. Godino, 2010Concimare le foglie per aumentare la resa. Olivo&Olio, 34043

122. Von, 1998State of commercial use of AMF-inoculum in Germany. In: Gianinazzi, S.; Schuepp, H. (eds.) Arbuscular mycorrhizas in sustainable soil-plant systems. Report of 1997 activities. Iceland: Cost Action 8.21, 153

123. E. Wollny, 1898Untersuchungen über den Einflub der mechanischen Bearbeitung auf die Fruchtbarkeit des Bodens. Forsch. Geb. Agrik. Phys. 20231290

124. Xiloyannis, A. M. Palese, 2001Efficienza dell'uso dell'acqua nella coltivazione dell'olivo. In COI; Regione Campania; CNR Ist. Irrigazione (Eds) "Gestione dell'acqua e del territorio per un olivicoltura sostenibile", Corso Internazionale di aggiornamento tecnico scientifico, Napoli, Italy.

125. Xiloyannis, B. Dichio, 2006L'irrigazione sostenibile in frutticoltura. Ital. J. Agron. 3507517

Chapter 2

Heavy metals bioaccumulation in Berseem (*Trifolium alexandrinum*) cultivated in areas under intensive agriculture, Punjab, India

Sandip Singh Bhatti[1], Vasudha Sambyal[2] and Avinash Kaur Nagpal[1]

[1]Department of Botanical and Environmental Sciences, Guru Nanak Dev University, Amritsar, India.

[2]Department of Human Genetics, Guru Nanak Dev University, Amritsar, India.

ABSTRACT

Berseem (*Trifolium alexandrinum*) is one of the main fodder crops of Punjab, India. But due to the heavy metal contamination of agricultural soils by anthropogenic activities, there is rise in metal bioaccumulation in crops like Berseem. In addition to human influence, heavy metal contents in soil are highly dependent on soil characteristics also. Therefore a study was conducted in areas having intensive agricultural practices to analyze physico-chemical characteristics of soils under Berseem cultivation and heavy metal bioaccumulation in Berseem. The studied soils were alkaline, sandy in texture and low in soil organic matter. Among the studied heavy metals (Cr, Cu, Cd, Co and Pb) in soil and Berseem, Cr content in Berseem was found to be above maximum permissible limits. Soil to Berseem metal bioaccmulation factor (BAF) was above 1 for Cr, Cu, Cd and Co in many samples and highest BAF was found for Co (4.625). Hence it can be concluded that Berseem from studied areas was unsafe for animal consumption.

KEYWORDS

Berseem, Heavy metals, Maximum permissible limits, Bioaccumulation factor

1. BACKGROUND

Punjab is one of the most fertile states of India and agriculture in addition to livestock rearing is the main source of income of people of this region. Berseem (*Trifolium alexandrinum*) is the main fodder crop of Rabi season in Punjab. Due to presence of perennial rivers Beas and Sutlej and abundant groundwater resources intensive agriculture is practised in Punjab throughout the year. Berseem is a leguminous crop belonging to family Fabaceae. It is fast growing, high biomass yielding fodder crop and is highly liked by the animals (Ali et al. 2012; Bhat 2013). But rise in the level of environmental pollutants in Berseem due to uncontrolled use of agrochemicals and polluted irrigation water is a serious cause of concern for livestock production systems (Rajaganpathy et al. 2011).

Heavy metals are one of the main pollutants which affect the plants and animals throughout globe. Food and fodder crops raised on metal contaminated soils have the tendency to accumulate excessive amounts of heavy metals, which poses severe risk to human and animal health (Rattan et al. 2005; Kulhari et al. 2013). Heavy metal contents in fodder crops are dependent on soil, climatic factors, agrochemical application, irrigation water quality, plant growth rates and plant parts (Ahmad et al. 2011). The heavy metal contents in agricultural soil are affected by cropping practices and soil properties (Dheri et al. 2007; Endalamaw and Chandravanshi 2015). In the areas which are situated on the banks of rivers there is an added risk of heavy metal contamination in soils due to rise in pollution in rivers, because in these areas river water is also used for irrigation. Although heavy metals like Co, Cr and Cu are essential for plant and animal metabolism, at levels above maximum permissible limits they disrupts the normal functioning of organisms (Ali et al. 2012; Jolly et al. 2013). Cd and

Pb are known to be highly toxic and carcinogenic for animals and humans (Rajaganpathy et al. 2011). Since heavy metal contents in soil are highly affected by soil characteristics and heavy metal bioaccumulation in fodder crops is a major route of heavy metal entry into livestock systems, a study was conducted in intensively cultivated areas situated on banks of rivers Sutlej and Beas to assess physcio-chemical characteristics of soil under Berseem cultivation and heavy metal bioaccumulation in Berseem.

2. RESULTS AND DISCUSSION

2.1 Soil physico-chemical characteristics

The physico-chemical characteristics of the studied soils are given in Table 1. The Pearson's correlation matrix of physico-chemical properties and heavy metals in studied soils are given in Table 2. The soils were found to be slightly alkaline in nature and the pH ranged from 7.38 to 7.99. Soil conductivity which indicates the salinity was maximum at site I (0.723 mS/cm) and minimum at site II (0.307 mS/cm). The observed soils were found to be sandy in texture with sand contents ranging from 3.15 (Site I) to 5.33 % (Site II). The sandy texture of soils is mainly due to settling of sandy river alluvium deposited for thousands of years by rivers Sutlej and Beas. The soil organic matter (SOM) which is one of the most important indicators of soil health (Rattan et al. 2005) was found to be near desertification levels (1.83–2.91 %) in the present study. The main reason for such low levels of SOM is the poor silt and clay contents of soils (Sollins et al. 1996). Significant positive correlation was observed between soil pH and SOM, which was in line with the observations of Aciego Pietri and Brookes (2008), but was contradictory to the trend observed by Yali et al. (2012). The carbonate content ($CaCO_3$) ranged from 5.07 to 9.04 % revealing the calcareous nature of the studied soils. The calcium (Ca) and magnesium (Mg) contents in soils ranged from 0.49 to 1.00 meq/100 g and 0.73–1.74 meq/100 g respectively. Ca and Mg are secondary nutrients for plants and abundantly available in soils to meet plant requirements. But excessive amount of Ca in soils sometimes causes deficiency of other plant nutrients (Troeh and Thompson 2005), which is evident from the negative correlation of Ca with sodium (Na), potassium (K) and nitrogen (N) (Table 2). The major soil nutrients nitrogen (N), phosphorous (P) and potassium (K) ranged from 114.33 to 665.15, 49.99–384.91 and 1052.83–1559.67 mg/kg respectively. The soil nutrients (Na, N and K) were found to be negatively correlated to sand and positively correlated to clay (Table 2), which can be due to high affinity of clay to soil nutrients (Boluda et al. 2011). The main anthropogenic sources of soil nutrients in the studied soils are NPK fertilizers, which are extensively used in the studied area. The NPK fertilizers are favored by the high pH of soils (Gil et al. 2004), which is evident from the positive correlation of pH with N, P and K (Table 2). The levels of soil physico-chemical parameters in the present study were similar to the levels observed in other parts of Punjab (Dheri et al. 2007; Katnoria et al. 2011).

2.2 Heavy metal contents in soil

Figure 1 shows the contents of different heavy metals (Cr, Cu, Cd, Co and Pb) studied in soil and Berseem. Among these metals, Cr was found to be highest in all soil samples. Maximum amount of Cr was observed at Site I (75.70 mg/kg) and minimum level was observed at Site III (21.37 mg/kg). Cu, Cd and Co ranged from 13.67 to 25.0, 0.15–0.37 and 0.13–3.83 mg/kg respectively. Pb contents ranged from 2.83 mg/kg (Site I) to 9.17 mg/kg (Site IV). The main sources of heavy metals in the soil samples are parent rock material, polluted irrigation water and various agrochemicals (fertilizers, pesticides, weedicides etc.) used for cultivation. The studied sites are located on the banks of rivers Sutlej and Beas which have been shown to be highly polluted with industrial effluents, sewage discharge and agricultural runoffs (Kaur et al. 2014a, b). The groundwater is also contaminated due to leaching of pollutants from surface water and upper soils layers. Some earlier studies have reported dangerous levels of heavy metals in the surface and groundwater of neighbouring regions of the sampled areas and other parts of Punjab (Kumar et al. 2007; Bhalla et al. 2011; Singh et al. 2013; Kaur et al. 2014a, b; Shrivastava 2014). Sewage discharge and effluents originating from various industries such as electroplating, dying, leather tanning, alloys, paints etc. situated upstream of these sites, especially in the neighbouring cities Ludhiana, Jalandhar, Kapurthala etc. are the main source of contamination of surface and ground water in these areas. The polluted irrigation water acts as a major source of heavy metals in the analyzed soil samples. Secondly, excessive use of NPK fertilizers and other agrochemicals further contaminate the soil with heavy metals. Various NPK fertilizers act as source of heavy metals such as Cd, As, Pb, Cr, Ni, Cu etc. The Phosphate fertilizers in particular, have substantial amounts of Cd in them (Mortvedt 1996; Milinovic et al. 2008; Savci 2012). Among the studied heavy metals in the present work Cu, Co and Cr are required in small amounts for plant metabolism. Cd and

Pb are non essential and toxic elements for plants and other soil organisms (Mertz 1981), which was evident from the negative correlation observed between SOM, Pb and Cd (Table 2). Pb being a toxic metal disrupts the total chlorophyll content of plants and decreases root expansion by restricting the cell division and cell elongation (Eun et al. 2000; McDermott et al. 2011). Thus it is a restricting factor for SOM.

The heavy metal contents in soil are also dependent on soil physico-chemical properties, which affect the mobility, availability and ecotoxicological risks of heavy metals (Boluda et al. 2011). The heavy metal contents in soil samples observed in the present analysis are below the maximum permissible limits (Table 3) despite the extensive agricultural practices and polluted irrigation water, which can be attributed to sandy texture of soil. Due to high sand contents in soil the metals can leach to lower soil layers (Boluda et al. 2011), which was evident from the negative correlation between heavy metals and sand (Table 2). The contents of Pb and Cd observed in the present study were lower and Cr contents were higher than the earlier works done by Dheri et al. (2007) and Chahal et al. (2014).

Table 1. Physico-chemical characteristics of soil under Berseem cultivation

Physico-chemical properties	Sites			
	Site I	Site II	Site III	Site IV
pH	7.99 ± 0.25	7.60 ± 0.71	7.38 ± 0.94	7.88 ± 0.74
Conductivity (mS/cm)	0.723 ± 0.066	0.307 ± 0.038	0.617 ± 0.089	0.440 ± 0.057
Sand (%)	88.33 ± 3.36	76.67 ± 9.43	90.33 ± 3.32	85.33 ± 5.31
Silt (%)	8.33 ± 1.64	18.0 ± 2.14	6.33 ± 1.66	10.33 ± 2.31
Clay (%)	3.15 ± 0.66	5.33 ± 1.33	3.66 ± 0.97	4.33 ± 1.01
SOM (%)	2.91 ± 0.89	1.83 ± 0.31	1.88 ± 0.35	2.68 ± 0.21
$CaCO_3$ (%)	5.07 ± 0.33	8.61 ± 0.97	9.04 ± 1.16	5.70 ± 0.78
Ca (meq/100 g soil)	0.87 ± 0.13	0.67 ± 0.06	1.00 ± 0.18	0.49 ± 0.07
Mg (meq/100 g soil)	1.74 ± 0.17	0.94 ± 0.14	0.73 ± 0.09	0.79 ± 0.13
Na (mg/kg)	354.33 ± 11.57	403.50 ± 35.42	404.16 ± 17.76	1073.67 ± 98.92
N (mg/kg)	319.67 ± 76.56	665.15 ± 140.93	114.33 ± 16.65	254.33 ± 42.77
P (mg/kg)	384.91 ± 54.46	49.99 ± 11.57	190.44 ± 27.01	98.15 ± 14.27
K (mg/kg)	1052.83 ± 81.79	1315.72 ± 71.98	1183.33 ± 87.66	1559.67 ± 101.03

SOM, soil organic matter, *$CaCO_3$* Carbonates, *Ca* calcium, *Mg* magnesium, *Na* sodium, *N* Kjehldal nitrogen, *P* available phosphorous, *K* potassium

Table 2. Pearson's correlation matrix of physico-chemical properties and heavy metal contents of soil

	pH	Cond.	Sand	Silt	Clay	SOM	Ca	Mg	Na	K	N	P	$CaCO_3$	Cr	Cu	Cd	Co
Cond.	0.23																
Sand	0.03	0.88**															
Silt	−0.02	−0.85**	−0.99**														
Clay	−0.05	−0.83**	−0.85**	0.79**													
SOM	0.62*	0.39	0.32	−0.32	−0.24												
Ca	−0.39	0.51	0.44	−0.42	−0.47	−0.08											
Mg	0.64*	0.53	0.15	−0.11	−0.34	0.39	0.17										
Na	0.33	−0.34	−0.01	−0.04	0.18	0.30	−0.60*	0.89**									
K	0.06	−0.71**	−0.38	0.33	0.52	0.01	−0.65*	−0.60*	0.89**								
N	0.17	−0.61**	−0.86**	0.88**	0.61*	−0.37	−0.33	0.19	−0.21	0.08							
P	0.43	0.90**	0.64*	−0.60*	−0.68*	0.49	0.43	0.72**	−0.42	0.75**	−0.33						
$CaCO_3$	−0.92**	−0.38	−0.26	0.25	0.27	−0.77*	0.23	−0.57	−0.40	−0.04	0.12	−0.51					
Cr	0.13	0.13	−0.13	0.09	0.28	−0.24	0.00	0.24	−0.43	−0.42	0.30	0.20	0.01				
Cu	−0.17	0.66	0.80**	−0.77**	−0.83**	0.43	0.54	0.02	−0.04	−0.30	−0.81**	0.48	−0.13	−0.52			
Cd	−0.18	0.14	−0.19	0.20	0.10	−0.49	0.07	0.27	−0.77**	−0.64*	0.33	0.25	0.34	0.69*	−0.35		
Co	0.54	−0.23	0.02	−0.05	0.14	0.45	−0.65*	−0.20	0.97**	0.80**	−0.17	−0.25	−0.59*	−0.38	−0.04	−0.73**	
Pb	−0.10	−0.54	−0.12	0.08	0.29	−0.02	−0.59*	−0.70*	0.89**	0.93**	−0.20	−0.67*	0.03	−0.49	−0.05	−0.65*	0.77**

* Correlation is significant at the 0.05 level (2-tailed)

** Correlation is significant at the 0.01 level (2-tailed)

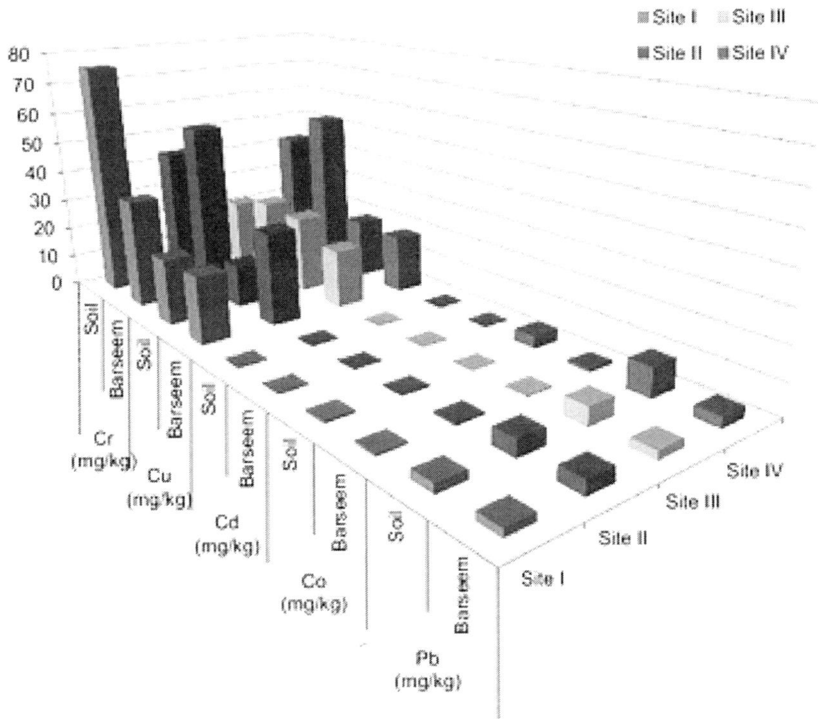

Figure1. Heavy metal contents in soil and Berseem samples of Punjab, India

Table 3. National and international maximum permissible limits of heavy metals for soil and fodder

	Chromium (mg/kg)	Copper (mg/kg)	Cadmium (mg/kg)	Cobalt (mg/kg)	Lead (mg/kg)
Maximum permissible limits for soil					
Indian (Awashthi 2000)*	–	135–270	3–6	–	250–500
European Union (2002)	150	140	3.0	–	300
Maximum permissible limits for fodder					
CERSPC (2009)	10.0	–	0.5	–	5.0

* Prevention of Food Adulteration Act No. 37 of 1954. Central and State Rules as Amended for 1999 (Sharma et al.2006; Singh et al. 2010)

2.3 Heavy metal contents in Berseem

Similar to the results observed in soil, Cr contents were highest in Berseem in comparison to the other metals (Fig. 1). Maximum amount of Cr was found in Berseem samples from Site II (43.43 mg/kg) and minimum content was observed at Site III (25.92 mg/kg). Cr contents were alarmingly higher than the maximum permissible limits for fodder set by Tolerance Limit of Heavy Metals for Feed of China (CERSPC 2009) in all the samples (Table 3). Cr content above the permissible limits can cause deleterious effects on plant physiological processes such as photosynthesis and respiration (Ahmad et al.2013). Although trace amounts of Cr is required for glucose metabolism in animals, but Cr overdose can cause liver necrosis, nephrites, gastrointestinal irritation and ulcers (coetaneous, nasal and mucous membrane) (Edwards and Gregory 1991; ATSDR 2000). The contents of Cu and Cd ranged from 13.67 to 21.35 and 0.15–0.37 mg/kg respectively. Cu is an important element for plant nutrition but Cd despite being a non-essential element in plants gets highly accumulated in plants (Nadian 2004). Excess amount of Cd in animal fodder can result in renal tubule damage, cardiovascular diseases, cancer, ostomalacia and deleterious effects on calcium, phosphorous and bone metabolism (Katole et al. 2013). Co contents were found to be maximum at Site IV (1.07 mg/kg) and minimum at Site III (0.62 mg/kg) which can be due to minimum and maximum levels of Co at the same sites in soil samples. Pb contents ranged from 2.5 to 4.33 mg/kg which is found to be close to the maximum permissible limit of 5 mg/kg (Table 3) at Site II (4.33 mg/kg). Pb at high levels decreases root expansion by restricting cell division and elongation in plants. Excess amounts of Pb in animal fodder can cause acute or chronic poisoning in animals leading to decreased haemoglobin synthesis, neurobehavioral impairment, peripheral neuropathy reproductive effects and neurotoxic malfunctioning in infants (Allcroft and Blaxter 1950). In Punjab lower levels of Cr and Cd and higher level of Pb in Berseem were observed by Dheri et al. (2007). Very high levels of Pb (102–383 mg/kg) were observed in fodder feed to animals in industrial areas of Punjab by Sidhu et al. (1994) which resulted in Pb toxicity (blood) in bovine animals. Heavy metal toxicity (Pb, Cd, Cr, Cu, Mn etc.) in animals due to contaminated fodder has been observed by many researchers in other parts of India also (Dey et al. 2011; Gowda et al. 2003; Raj et al. 2006). Therefore the Berseem fodder from the studied areas was contaminated with heavy metals due to combined effect of several anthropogenic factors (such as pollution of irrigation water, excessive use of agrochemicals, etc.) and is a serious cause of concern for animal health.

2.4 Metal bioaccumulation factor (BAF)

Heavy metal bioaccumulation factor (BAF) values above 1 were observed for Cr, Cu, Cd and Co (Fig. 2) in many samples, which showed high level of metal bioaccumulation in Berseem. Highest value of BAF was observed for Co (4.625) at Site III. Studies on metal uptake by plants have revealed that heavy metals are passively transported from roots to shoots via xylem vessels and are preferably accumulated in areas with high transpiration rates (Tamoutsidis et al. 2009). Since the most important parts in Berseem from commercial point are leaves and shoot, heavy metal accumulation in these parts poses severe risk to animals. This problem is further aggravated by the high phytoremediation potential of Berseem (Ali et al. 2012; Bhat 2013).

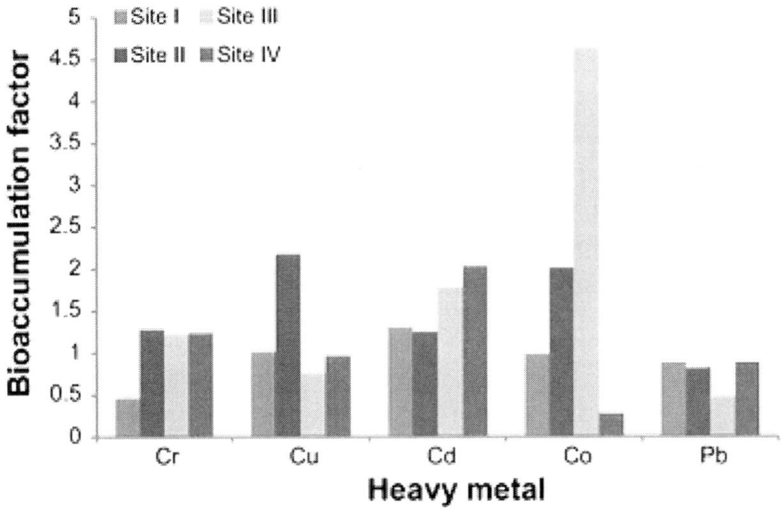

Figure 2. Soil to Berseem heavy metal bioaccumulation factor (BAF)

3. CONCLUSIONS

The present study revealed that the areas studied have alkaline soils with sandy texture, calcareous nature and low SOM. The Cu, Cd, Co and Pb contents in studied soil and Berseem samples were lower than the maximum permissible limits. But Cr contents in Berseem were higher than the maximum permissible limits for fodder (CERSPC 2009). BAF was found to be above 1 for Cr, Cu, Cd and Pb for many Berseem samples. Therefore it can be concluded that there is significant bioaccumulation of heavy metals in Berseem grown in these intensively cultivated areas which can be due to synergistic effect of various anthropogenic factors such as irrigation water contamination, indiscriminate use of agrochemicals, industrial activities, etc. and thus it is unsafe for animal consumption.

4. METHODS

4.1 Study area

The state of Punjab (Lat. 29°30–32°32′N and Long. 73°55 and 76°50′E) is located in the north-western part of India bordering Pakistan. The annual rainfall in Punjab is 435.6 mm and has a continental, semiarid to subhumid climate with two main crop seasons Kharif (fall) and Rabi (spring). Four sites were selected for sampling, which are situated on the banks of rivers Beas and Sutlej. The geographical coordinates of sites are given in Table 4 and Fig. 3 shows the map of studied area. The upstream regions of the sites chosen for sampling are receiving effluents from various industries (electroplating, dying, leather tanning, alloys, paints etc.) and sewage. The main occupation in the study area is agriculture which includes extensive use of agrochemicals (fertilizers, pesticides and weedicides).

Table 4. List of sampling sites with their districts, adjoining river, geographical coordinates

S. No.	Sites	Village	District	Adjoining river	Coordinates
1.	Site I	Rajewal	Kapurthala	Beas	N31°22′59.2″ E075°10′47.6″
2.	Site II	Yousufpur	Kapurthala	Sutlej	N31°08′47.1″ E075°06′22.4″
3.	Site III	Tibbi Taiba	Firozpur	Sutlej	N31°07′23.1″ E075°05′42.4″
4.	Site IV	Doomniwala	Tarn Taran	Sutlej	N31°09′43.1″ E074°48′50.0″

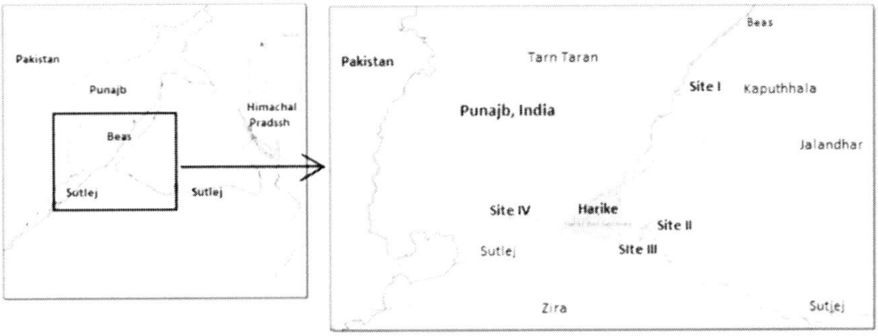

Figure 3. Soil and Berseem sampling sites in Punjab, India

4.2 Sampling and preparation

Soil sampling was done during the period of March–April 2013. From each site composite soil samples in triplicates were collected from fields under Berseem cultivation. At least five subsamples of soil were pooled to form a composite sample. Soil samples were taken from depths of 0–15 cm. Composite samples of Berseem (Shoots) were collected in triplicates from corresponding soil sampling fields from each site for heavy metal analysis. All soil and Berseem samples were stored in clean polythene bags and were brought to the laboratory. The soil samples were air-dried, ground and passed through 2 mm sieve for physico-chemical and heavy metal analysis. The Berseem shoot samples were washed with deionised water, oven dried at 70 °C and then grounded to fine powder with pestle mortar. Similar methods have been used for collection and preservation of soil and plant samples by researchers earlier also (Yao et al. 2010; Bermudez et al. 2011; Rodriguez Martin et al. 2013; Liu et al. 2013).

4.3 Physico-chemical analysis

The soil pH and conductivity were determined in 1:5 soil:water suspension using HM digital meter-COM-100 (New Delhi, India) and Equip-tronics EQ-614-A (Mumbai, India), respectively. The soil suspension was prepared by mixing soil and water in desired proportion (1:5) and the mixture was shaken for 2 h and the supernatant was filtered and used for measurement. Soil texture and organic carbon content was determined by Hydrometer method (Jacob and Clarke 2002) and Walkley–Black wet oxidation method

(Nelson and Sommers 1982) respectively. A factor of 1.72 was multiplied with organic carbon content to determine SOM. EDTA titration method was used for measuring calcium (Ca) and magnesium (Mg) (Lanyon and Heald 1982), acid neutralization method for $CaCO_3$ (Hesse 1971) and potassium (K) and sodium (Na) were measured by using a Systronics Flame Photometer-128, after digesting the samples in a diacid mixture (HClO4/HNO3 in a 4:1 ratio) (Bhat et al. 2014). Total nitrogen (N) was determined by Kjeldahl method (Bremner and Mulvaney 1982) and available phosphorous (P) by sodium bicarbonate extraction method using Spectrophotometer-2202 (Ahmedabad, Gujarat, India) (Olsen et al. 1954).

4.4 Heavy metal analysis

For heavy metal (Cr, Cu, Cd, Co and Pb) determination one gram of soil was digested with 15 mL of aqua regia (HNO_3: HCl in 3:1 ratio) and 1 g Berseem shoot sample with 15 mL of triacid mixture (HNO_3:H_2SO_4:$HClO_4$ in 5:1:1 ratio) at 80 °C till a transparent solution was obtained (Allen et al. 1986). The digested samples were filtered and diluted with de-ionized water up to 50 mL and analyzed for the metals viz. Chromium (Cr), copper (Cu), cadmium (Cd) cobalt (Co) and lead (Pb) by flame atomic absorption spectrophotometer (AAS) (Agilent 240 FS AA model). Properly washed glassware, double distilled water and analytical grade reagents were used throughout the study. The standard solutions of selected heavy metals were procured from Agilent (1000 mg/L) and were used to make solutions of varying concentrations by dilution of the standards. After every ten sample readings, the standards were run to assure the working of machine with 95 % accuracy (Arora et al. 2008).

4.5 Metal bioaccumulation factor (BAF)

Heavy metal accumulation of soil and Berseem were calculated on the basis of dry weight. The metal bioaccumulation factor is a ratio of heavy metal concentration of crop to soil (Zhuang et al. 2013) and was calculated as follows:

$$BAF = C_{plant}/C_{soil} \tag{1}$$

where C_{plant} and C_{soil} are the concentrations of heavy metal in Berseem and soil, respectively, on a dry weight basis.

4.6 Statistical analysis

The analysis of physico-chemical characteristics and heavy metal content of soil and Berseem samples was done in triplicates and the data is presented as mean ± standard error. Pearson correlation coefficients were calculated to analyze the correlation between physico-chemical properties and heavy metals in soil. Statistical analysis was done with the help of IBM SPSS version 16.0 (Chicago, USA) and Microsoft excel computer software programs.

AUTHORS' CONTRIBUTIONS

SSB was involved in design of sampling, the chemical analysis and in the initial drafting of the manuscript. VS was involved in revising the manuscript critically for important intellectual content. AVN gave substantial contributions to conception and design, acquisition interpretation of data and overall supervision. All authors read and approved the final manuscript.

ACKNOWLEDGEMENTS

The authors acknowledge UGC (UPE and RGNF) for financial support and Head, Department of Botanical and Environmental Sciences, Guru Nanak Dev University, Amritsar for laboratory facilities.

REFERENCES

1. Aciego Pietri JC, Brookes PC (2008) Relationships between soil pH and microbial properties in a UK arable soil. Soil Biol Biochem 40:1856–1861

2. Ahmad K, Ibrahim M, Khan ZI, Rizwan Y, Ejaz A, Fardsous A, Gondal S, Lee DJ, Al-Yemeni M (2011) Effect of sewage water on mineral nutritive potential of six fodder species grown under semiarid conditions. Saudi J Biol Sci 18:317–321

3. Ahmad K, Shaheen M, Khan ZI, Bashir H (2013) Heavy metals contamination of soil and fodder: a possible risk to livestock. Sci Tech Dev 32(2):140–148

4. Ali H, Naseer M, Sajad MA (2012) Phytoremediation of heavy metals by *Trifolium alexandrinum*. Int J Environ Sci 2(3):1459–1469

5. Allcroft R, Blaxter KL (1950) Lead as a nutritional hazard to farm livestock. The toxicity of lead to cattle and sheep and an evaluation of the lead hazard under farm conditions. J Comp Pathol 60(3):209–218

6. Allen SE, Grimshaw HM, Rowland AP (1986) Chemical analysis. In: Moore PD, Chapman SB (eds) Methods in plant ecology. Blackwell, Scientific Publication, Oxford, London, pp 285–344

7. Arora M, Kiran B, Rani S, Rani A, Kaur B, Mittal N (2008) Heavy metal accumulation in vegetables irrigated with water from different sources. Food Chem 111:811–815

8. ATSDR (2000) Toxicological profile of chromium. Agency for Toxic Substances and Drug Registry, U S Department of Health and Human Services, Atlanta, GA

9. Awashthi SK (Ed.) (2000). Prevention of Food Adulteration Act No. 37 of 1954. Central and State Rules as Amended for 1999, Ashoka Law House, New Delhi

10. Bermudez GMA, Jasan R, Pla R, Pignata ML (2011) Heavy metal and trace element concentrations in wheat grains: assessment of potential non-carcinogenic health hazard through their consumption. J Hazard Mater 193:264–271

11. Bhalla A, Singh G, Kumar S, Shahi JS, Mehta D (2011) Elemental analysis of ground water from different regions of Punjab state (India) using EDXRF technique and the sources of water contamination. Int Conf Environ Comput Sci 19:156–164

12. Bhat S (2013) Phytoremidation properties and CLA content of Berseem (*Trifolium alexandrinum*). Asian J Microbiol Biotechnol Environ Sci 15(3):573–577

13. Bhat SA, Singh J, Vig AP (2014) Genotoxic assessment and optimization of pressmud with the help of exotic earthworm *Eisenia fetida*. Environ Sci Pollut Res 21(3):8112–8123

14. Boluda R, Roca-Perez L, Marimon L (2011) Soil plate assay: an effective method to determine ecotoxicological risks. Chemosphere 84:1–8

15. Bremner JM, Mulvaney CS (1982) Nitrogen total. In: Page AL, Miller RH, Keeney DR (eds) Methods of soil analysis. American Society of Agronomy, Madison, WI, pp 575–624

16. Chahal V, Chand P, Nagpal A, Katnoria JK, Pakade YB (2014) Evaluation of heavy metals contamination and its genotoxicity in agricultural soil of Amritsar, Punjab, India. Int J Res Chem Environ 4(4):20–28

17. Chief Editor Room of Standard Press of China (CERSPC) (2009) Compilation of standards for feed industry. Standard Press of China, Beijing **(in Chinese)**

18. Dey S, Dwivedi SK, Swarup D (2011) Heavy metal contaminants in soil, water and fodder and their presence in livestock and products: a review. J Environ Sci Technol 4(3):234–249

19. Dheri GS, Brar MS, Malhi SS (2007) Heavy-metal concentration of sewage-contaminated water and its impact on underground water, soil, and crop plants in alluvial soils of northwestern India. Commun Soil Sci Plant Anal 38:1353–1370

20. Edwards WC, Gregory DG (1991) Livestock poisoning from oil field drilling fluids, muds and additives. Vet Hum Toxicol 33(5):502–504

21. Endalamaw FD, Chandravanshi BS (2015) Levels of major and trace elements in fenel (*Foeniculum vulgari* Mill.) fruits cultivated in Ethiopia. Springerplus 4:5

22. Eun SO, Youn HS, Lee Y (2000) Lead disturbs microtubule organization in root meristem of *Zea mays*. Physiol Plant 110:357–365

23. European Union (2002) Heavy metals in wastes, European Commission on Environmet. http://ec.europa.eu/environment/waste/studies/pdf/heavymetalsreport.pdf. Assessed 21 July 2014

24. Gil C, Boluda R, Ramos J (2004) Determination and evaluation of cadmium, lead and nickel in greenhouse soils of Almería (Spain). Chemosphere 55(7):1027–1034

25. Gowda NKS, Malathi VS, Jash S, Roy KS (2003) Status of pollutants and trace elements in water, soil, vegetation and dairy animals in industrial area of Banglore. Indian J Dairy Sci 56:86–90

26. Hesse PR (1971) A textbook of soil chemical analysis. John Murray, London, p 52

27. Jacob H, Clarke G (2002) Part 4, Physical Method. Soil Science Society of America, Methods of Soil Analysis, Madison, Wisconsin, USA, p 1692

28. Jolly Y, Islam A, Akbar S (2013) Transfer of metals from soil to vegetables and possible health risk assessment. Springerplus 2:385

29. Katnoria JK, Arora S, Bhardwaj R, Nagpal A (2011) Evaluation of genotoxic potential of industrial waste contaminated soil extracts of Amritsar, India. J Environ Biol 32:363–367

30. Katole SB, Kumar P, Patil RD (2013) Environmental pollutants and livestock health: a review. Vet Res Int 1(1):1–13

31. Kaur J, Chaudhary A, Kaur R (2014a) Assessment of mutagenic, genotoxic and cytotoxic potential of water samples of Harike wetland: a Ramsar site in India using different ex vivo biological systems. Ecotoxicology 23:967–977

32. Kaur T, Sharma K, Sinha AK (2014b) Industrial pollution in the sub-soil water and its health effects: a preliminary study around Buddha Nullah, Punjab. In: Sharma K, Sinha AK (eds) Human ecology in an era of globalization and urbanization. Serial Publications, New Delhi, pp 267–282

33. Kulhari A, Sheorayan A, Bajar S, Sarkar S, Chaudhury A, Kalia RK (2013) Investigation of heavy metals in frequently utilized medicinal plants collected from environmentally diverse locations of north western India. Springerplus 2:676

34. Kumar M, Kumari K, Ramanathan AL, Saxena R (2007) A comparative evaluation of groundwater suitability for irrigation and drinking purposes in two intensively cultivated districts of Punjab, India. Environ Geol 53:553–574

35. Lanyon LE, Heald WR (1982) Magnesium, calcium, strontium and barium. Agronomy No. 9

36. Liu X, Song Q, Tang Y, Li W, Xu J, Wu J, Wang F, Brookes PC (2013) Human health risk assessment of heavy metals in soil-vegetable system: a multi-medium analysis. Sci Total Environ 463–464:530–540

37. McDermott S, Wu J, Cai B, Lawson A, Marjorie Aelion C (2011) Probability of intellectual disability is associated with soil concentrations of arsenic and lead. Chemosphere 84(1):31–38

38. Mertz W (1981) The essential trace elements. Science 213:1332–1338

39. Milinović J, Lukić V, Nikolic-Mandić S, Stojanović D (2008) Concentrations of heavy metals in NPK fertilizers imported in Serbia. Pestic Phytomed 23:195–200

40. Mortvedt JJ (1996) Heavy metal contaminants in inorganic and organic fertilizers. Fertil Res 43:55–61

41. Nadian H (2004) Cd and Mn uptake and bioaccumulation in *Trifolium alexandrinum* L.: interaction with mycorrhizal colonization. In: Proceedings of the fourth international Iran and Russia conference, pp 595–601

42. Nelson DW, Sommers LE (1982) Total carbon and organic matter. In: Page AL (ed) Methods of soil analysis, part 2. ASA-SSSA, Madison, WI

43. Olsen SR, Cole CV, Watanabe FS, Dean LA (1954) Estimation of available phosphorous in soils by extraction with sodium bicarbonate. USDA Circ. No. 939. Washington, DC, US Dept. Agric

44. Raj BG, Patnaik MC, Babu SP, Kalakumar B, Singh MV, Shylaja J (2006) Heavy metal contaminats in water-soil-plant-animal continuum due to pollution of Musi river around Hyderabad in India. Indian J Anim Sci 76:131–133

45. Rajaganpathy V, Xavier F, Sreekumar D, Mandal PK (2011) Heavy metal contamination in soil, water and fodder and their presence in livestock and products: a review. J Environ Sci Technol 4(3):234–249

46. Rattan RK, Datta SP, Chhonkar PK, Suribabu K, Singh AK (2005) Long-term impact of irrigation with sewage effluents on heavy metal content in soils, crops and groundwater: a case study. Agric Ecosyst Environ 109:310–322

47. Rodriguez Martin JA, Ramos-Miras JJ, Boluda R, Gil C (2013) Spatial relations of heavy metals in arable and greenhouse soils of a Mediterranean environment region (Spain). Geoderma 200–201:180–188

48. Savci S (2012) An agricultural pollutant: chemical fertilizer. Int J Environ Sci Dev 3(1):77–80

49. Sharma RK, Agarwal M, Marshall F (2006) Heavy metal contamination in vegetables grown in wastewater irrigated areas of Varanasi, India. Bull Environ Contam Toxicol 77:312–318

50. Shrivastava BK (2014) Elevated uranium and toxic elements concentration in groundwater in Punjab state of India: extent of the problem and risk due to consumption of unsafe drinking water. Water Qual Expo Health.

51. Sidhu PK, SandhuBS Singh J, Kwatra MS (1994) Lead toxicosisin bovine due to industrial pollution in Punjab. Indian J Anim Sci 64:1359–1360

52. Singh A, Sharma RK, Agrawal M, Marshall FM (2010) Health risk assessment of heavy metals via dietary intake of foodstuffs from the wastewater irrigated site of a dry tropical area of India. Food Chem Toxicol 48(2):611–619

53. Singh CK, Rina K, Singh RP, Mukherjee S (2013) Geochemical characterization and heavy metal contamination of groundwater in Sutlej River Basin. Environ Earth Sci.

54. Sollins P, Homann P, Caldwell BA (1996) Stabilization and destabilization of soil organic matter: mechanisms and controls. Geoderma 74(1):65–105

55. Tamoutsidis E, Lazaridou M, Papadopoulos I, Spanos T, Papathanasiou F, Tamoutsidou M, Mitlianga P, Vasiliou G (2009) The effect of treated urban wastewater on soil properties, plant tissue composition and biomass productivity in berseem clover and corn. J Food Agric Environ 7(3&4):782–786

56. Troeh FR, Thompson LM (2005) Soil and soil fertility, 6th edn. Wiley India Pvt Ltd, New Delhi

57. Yali W, Zhonghao Z, Gangcai L (2012) Physico-chemical properties and enzyme activities of the arable soils in Lhasa, Tibet, China. J Mt Sci 9(4):558–569

58. Yao F, Changcun L, Jianjun M, Tingcheng Z (2010) Effects of plant types on physico-chemical properties of reclaimed mining soil in inner Mangolia, China. Chin Geogr Sci 20(4):309–317

59. Zhuang P, Li Z, Zou B, Xia H, Wang G (2013) Heavy metal contamination in soil and soyabean near the Dabaoshan mine, South China. Pedosphere 23(3):298–304

Chapter 3

Agrochemicals and Bacterial Diversity in Cultivated Tropical Soils

Joseph Addo Ampofo[1]

[1] Council for Scientific and Industrial Research, Water Research Institute, CSIR –WRI, Accra, Ghana

1. INTRODUCTION

1.1. General

Agrochemicals provide yield protection, prevent and cure crop disease, provide insulation to reduce energy use and provide countless other benefits that increase the standard of living for civilization as a whole. While the chemicals industry has made good progress reducing its overall environmental footprint, chemicals can also create a negative impact on human health and the environment when their production and use are not managed responsibly (Edwards, 1975; Tu, 1990; Zulalian, 1990).

1.2. Agriculture In Ghana

Urban food needs is increasing in developing countries, including Ghana (Figure 1), with growing populations. Increasingly vegetables are grown in urban and peri-urban areas to meet this demand. Agriculture has been Ghana's most important economic sector over the years, employing more than 60% of the national work force, mainly in the small landholders on formal and informal basis (Gerken*et al*, 2001) and accounting for about 40% of the total GDP and export earnings. Ghana has climatic zones that range from dry savanna to wet forest and run in east-west bands across the country.

Figure 1. Map of Ghana, its boundaries and Lake Volta. Also shown are some major rivers.

Agriculture crops including yams, grains, cocoa, oil palms, kola nuts and timber, form the base of Ghana's economy. Compared to this, the contribution of the more traditional vegetables such as okra, pepper, tomato, onion and egg plants to the agricultural GDP is low. However, considering their contribution to agricultural GDP and the land area devoted to the cultivation of all crops, the traditional vegetables produce more value per area. For example, the share in total area cultivated and value for vegetables (tomato, okra, pepper and eggplant) is 3.0 and 7.0%, respectively, while for cereals (maize, millet, sorghum and rice), these values are 35.0 and 8.0%, respectively (Table 1).

1.3. Application Of Agrochemicals In Agriculture

Vegetable production in Ghana typically occurs in intensely managed, irrigated smallholder farms with relatively high pesticide inputs. Surveyed vegetable crops (tomato, pepper, okra, eggplants (or garden eggs) and onion) cover approximately 0.4% of the cultured land of Ghana, equating to 58,270 ha in 1998 (Gerken *et al.*, 2001). An estimated average pesticide rate of 0.08 liters ai/ha is applied to these vegetables. Relatively lower quantities are applied to cereals (0.03 liters ai/ha1) and higher quantities to cocoa (0.5 liters ai/ha). The compounds applied in vegetable production include organochlorine and organophosphate insecticides. In contrast to the traditional organochlorines, organophosphates are not highly persistent, but some can be highly toxic to some aquatic organisms (Castillo *et al.*, 2006). Intensively managed vegetable farms are also characterized by an extensive network of drainage systems where surplus water may flow into local streams and rivers. Consequently, the aquatic ecosystems located downstream of vegetable farmlands might be vulnerable due to intensive pesticide use, drainage systems, and high precipitation rates typical for tropical areas where vegetable production occurs.

1.4. Non-Point Source Agricultural Pollution

Non-point source agricultural pollution is regarded as the greatest threat to the quality of surface waters in rural areas. One of the most important routes leading to non-point source agricultural pollution of surface waters in rural areas is runoff. Runoff from agricultural fields introduces pesticides, soil, organic matter, manure and fertilizer into small streams, increasing the volume of stream discharge and changing water quality (Neumann and Dudgeon, 2002). The impacts of such runoff are well documented (Cooper, 1993; Castillo *et al.*, 1997; DeLorenzo *et al.*, 2001).

Table 1. Growth rates and contributions of different crops to agricultural GDP

Crops	Share in value (%)	Share in area (%)	Rate of growth (%)	
			Area	Yield
Maize (*Zea mays*)	4.80	17.0	2.00	3.00
Rice (*Oryza glaberrina*)	1.00	2.23	8.00	2.00
Millet (*Pennisetum glaucum*)	1.30	7.27	2.00	-
Sorghum (*Sorghum bicolor*)	1.24	8.41	2.00	1.50
Cassava (*Manihot esculenta*)	19.2	12.1	2.00	3.00
Cocoyam (*Xanthosoma roseum*)	10.2	6.08	2.00	-
Yam (*Dioscorea* spp.)	16.7	6.32	2.36	0.75
Plantain (*Musa* spp.)	13.0	5.32	2.00	-
Groundnut (*Arachis hypogaea*)	3.83	4.66	2.36	3.00
Tomato (*Solanum lycopersicum*)	2.79	0.61	1.00	2.00
Okra (*Hibiscus esculentus*)	3.16	0.85	3.00	2.00
Pepper (*Capsicum annuum*)	0.14	1.71	3.00	2.00
Eggplant (*Solanum melongena*)	0.56	0.07	3.00	2.00
Beans (*Phaseolus vulgaris*)	15.3	5.12	1.00	2.00
Cocoa (*Theobroma cacao*)	0.92	18.8	5.00	2.00
Oil Palm (*Elaeis guineensis*)	0.07	1.58	2.00	2.00
Rubber (*Ficus elastic*)	0.06	0.06	4.00	2.00
Coffee (*Coffea Arabica*)	0.16	0.02	4.00	2.00
Cotton (*Gossypium hirsutum*)	0.12	0.32	8.00	2.00
Tobacco (*Nicotiana tabacum*)	3.47	0.06	8.00	2.00
Orange (*Citrus sinensis*)	0.09	0.55	3.00	2.00
Pineapple (*Ananas comusus*)	0.09	0.06	3.00	2.00
Total crops/growth rates	98.2%	99.2%	3.31%	1.58%

Source: Nurah, 1999

Traditional vegetable farming systems (i.e. without any chemical input) are incapable of meeting demand. For instance, pests and diseases, which pose big problems in vegetable production, require intensive pest management to control them. Chemical pesticide use is a common practice to control pests and diseases in vegetable cultivation. However, besides their beneficial effects, pesticides are accepted as having potential environmental and public health impacts as well. If improperly used, pesticides can cause direct human poisoning, accumulate as residues in food and the environment or lead to the development of resistant strains of pests. These problems can arise from misuse of the pesticides or over-reliance on them, particularly if the users are not aware of these potential problems.

A total of 43 pesticides are in use in vegetable farming in Ghana (Table 2). This figure was obtained as a direct summation of pesticides applied on farms, but it could be lower than the actual number of pesticides in use. The pesticides comprised insecticides, fungicides and herbicides. Herbicides (44%) were the class of pesticides most used in vegetable farming in the areas surveyed, followed by insecticides (33%) and fungicides (23%).

In Table 2 the classification of these pesticides by the type of pests they control, active ingredient, chemical group and World Health Organization (WHO) Hazard Category is presented. The herbicides and fungicides used are mostly under WHO Hazard Category III, with a few under Hazard Category II. All the insecticides used are under Hazard Category II, which WHO classifies as moderately hazardous. This category includes organochlorines (OCs), organophosphates (OPs) and pyrethroids. Endosulfan was the only OC mentioned in use in the survey.

Technical endosulfan, a mixture of two stereoisomers, that is, α- and β-endosulfan in the approximate ratio of 7:3 (Shetty *et al.*, 2000; Kennedy *et al.*, 2001), is a chlorinated pesticide for control of a large spectrum of insect pests on a wide range of crops (Aguilera-del Real *et al.*, 1997). It is used in many countries throughout the world for the control of pests on fruits, vegetables, tea, tobacco, and cotton (Antonious and Byers, 1997; Sethunathan *et al.*, 2002). Because of such abundant usage, and the potential for accumulation in the environment (endosulfan is not readily detoxified by soil microorganisms), residues are detectable in soils, sediments, and crops at harvest time (Goebel *et al.*, 1982; U.S. Department of Health and Human Services, 1990). Although the metabolites of endosulfan, that is., sulfate, diol, ether, hydroxy ether, and lactone, have been shown to occur (Maier-Bode, 1968;Schuphan *et al.*, 1968), only the sulfate metabolite is significant as a residue (Antonious and Byers, 1997).

Table 2. Types of Pesticides applied in vegetable production in Ghana

Pesticide type (% of total number in use)	Active Ingredient	Chemical Group	Hazard Category (WHO)	Registered for use on
Herbicide (44%)	Pendimethalin	Dinitroaniline	III	Tomatoes, onions
	2, 4-D	Aryloxyalkanoic acide	II	Rice, sugarcane
	Propanil	Anilide	III	Rice
	MCPA-Thioethy1	Aryloxyalkanoic acide	III	Not registered
	Oxadiazon	Oxadiazole	III	Not registered
	Oxyfluorfen	Diphenyl ether	III	Not registered
	Bensulfuron-methy1	Sulfonylurea	III	Rice
	Glyphosate	Glycine derivative	III	Various crops
	Paraquat dichloride	Bipyridylium	II	Various crops
	Acifluorfen	Diphenyl ether	III	Not registered
	Metolachlor	Chloroacetamide	III	Not registered
	Phenmedipham	Carbamate	III	Not registered
Fungicide (23%)	Mancozeb	Carbamate	III	Mangoes, vegetables
	Matalaxyl-M	Acylalanine	II	Not registered
	Thiophanate-methyl	Benzimidazole	III	Various crops
	Carbendazim	Benzimidazole	III	Not registered
	Benomyl	Benzimidazole	III	Not registered
Insecticide (33%)	Lambda-cyhalothrin	Pyrethroid	II	Vegetables
	Chlorpyrifos	Organophosphorus	II	Citrus, public health
	Endosulan	Organochlorine	II	Cotton
	Dimethoate	Organophosphorus	II	Not registered
	Cypermethrin	Pyrethroid	II	Not registered
	Deltamethrin	Pyrethroid	II	Various crops

Although the impacts are complex and often unknown or sometimes open to debate, some negative effects are well documented, such as chemicals found in the environment that are persistent, bioaccumulative and or toxic (e.g. PCBs, dioxins) (Waiwright, 1978; Moorman, 1989). Concern has been raised about chemicals which interfere with the normal function of hormonal systems of human and animals (i.e. endocrine disrupters), and substances which impact on children's health (De Reuck, *et al*, 1979).

1.5. Activity Of Selected Agrochemicals

Glyphosate, the main component of Ceresate, is a non-selective, non-residual herbicide used against annual or biennial herbaceous monocotyledons, herbaceous dicotyledonous and perennial weeds. It is absorbed by foliage and transported through plant and is very effective on many deep-roots perennial species. It is metabolized or broken down by some plants, while other plants do not break it down. It is not usually absorbed from the soil by plants (Rueppel *et* al, 1977; McEwen & Stephenson, 1979;Eberbach & Douglas, 1983). Glyphosate remains unchanged in the soil for varying lengths of time, depending on soil texture and organic matter content. The half-life of glyphosate can range from 3 to 130 days.

Dimethoate (Cerox) is an organophosphorus insecticide with a contact and systemic action. Dimethoxon, an oxygen analogue metabolite of dimethoate, appears to play a dominant role in its toxicity for insects and mammals (Bohn, 1964; Koppel etal, 1986). Hydrolytic degradation is the main inactivating pathway of dimethoate in the environment. The half-life of dimethoate in different plants is between 2 and 5 days. Degradation in soil is dependent on the type of soil, temperature, moisture, and pH level.

Paraquat is a selective herbicide used to control most annual grasses and certain broadleaf weeds in field corn, potatoes, rice, cotton, soybeans, tobacco, peanuts and sunflowers. It is used in both pre-emergence and early post-emergence weed control (McErtenson, 1992). Paraquat is not subject to microbial degradation. Slight losses of paraquat can result from photodecomposition and volatilization. Its soil half-life is 90 days.

1.6. Microorganisms In Soil

Microorganisms present in soil include Actinomycetes, Fungi, Algae, Bacteria and Protozoa. Most organisms are found in the top layers of soil, usually the top 2-3 centimeters, since this is typically where most of the organic matter is located (Alexander, 1979). The organisms are usually concentrated close to root surfaces in the rhizosphere, within living and dead roots, on soil particles, or among aggregates of soil particles. The rhizosphere is the region of the soil that is immediately adjacent to and affected by plant roots. It is a dynamic region where interaction takes place between plants, soil, microorganisms, nutrients and water.

Microorganisms play a major role in the breakdown of pesticides in the soil. Many microbes are capable of utilizing pesticides as sources of carbon and most pesticides studied are attacked at one or more sites by microorganism e.g. the bacteria *Hydrogenomones* can degrade DDT completely to carbon dioxide (McEwen & Stephenson, 1979).

Edwards (1975) lists possible effects on living organisms in soil contaminated with insecticides to include (i) direct toxic effect to microbial life in the soil, (ii) affecting organisms genetically to produce populations resistant to pesticides, (iii) sub-lethal effects resulting in alterations in behavior or changes in metabolic or reproductive activities, and, (iv) absorption into the bodies of soil fauna and passing on to other organisms.

The study determined the effects of some selected agrochemicals on bacterial population in the soil, and investigated the effect of agrochemicals on plant growth.

2. MATERIALS AND METHODS

2.1. The Agrochemicals Used

The agrochemicals used were obtained from the open market. Each agrochemical was new and sealed in bottles of one litre volume. The products included (i) Cerox, an insecticide containing dimethoate 400g / L; (ii) Ceresate, a herbicide containing glyphosate IPA 4% w/w SL; and, (iii) Paraquat, a herbicide with composition, paraquat DCL 24% w/w SL.

2.2. Selection Of Viable Seeds For Planting

Undamaged bambara groundnut (*Vigna subterranea*) seeds of similar sizes were surface-sterilized by immersing for 5 minutes in 0.1% mercuric chloride (HgCl) solution and washed in six changes of sterile distilled water. The seeds were next washed in 70% ethanol for 3 minutes, and rinsed twice with sterile distilled water. The sterile seeds were placed on water agar (0.1% agar) in large Petri dishes and incubated at room temperature for 5 days. The vigorously germinating seeds were selected for planting.

2.3. Planting Of Seeds In Experimental Soil

Five seeds were sowed in each pot, and the seedlings were thinned to one after they had survived. There were four replicates for each soil treated type. The plants received full sunlight up to mid-day each day and were protected from rains. They were watered daily with 20ml tap water per pot. Once a week, each pot received, in addition, 10ml Sachs' solution to augment the nutrient content in the soil.

2.4. Application Of Agrochemicals

On the second week of planting, the agrochemicals were applied to the soil with the seedlings. Pots labelled A served as control, no agrochemical was applied. Pots labelled B were sprayed with Cerox. Pots labelled C were sprayed with Ceresate, and Pots labeled D were sprayed with Paraquat. Spraying was done using a spray bottle. The agrochemicals were diluted with sterile distilled water according to the manufacturer's recommendation as follows:

Ceresate:	5 ml:300 ml of distilled water
Cerox:	1.8 ml:300 ml of distilled water
Paraquat:	1.2 ml :300 ml of distilled water

2.5. Assessment Of Extent Of Growth Of Experimental Plants

The following records were made of the bambara groundnut plants: number of leaves, leaf length and leaf broadness. These measurements were taken once a week, for four weeks with the use of a ruler.

2.6. Assessment Of Nodulation

After six weeks, the plants were harvested and the roots thoroughly washed. The nodules were detached and counted.

2.7. Enumeration Of Total Heterotrophic Bacterial Populations In Soils

Soil samples were taken one day after planting of the seedlings and another set of samples on the third week after planting, from each of the pots to determine the bacterial population present in each soil treatment type. The soil samples were taken from between 2 - 8cm away from the stem of each seedling. One gram of each soil sample was dissolved in 9ml of sterile saline water and thoroughly mixed. Serial dilutions were made of each solution and 1ml plated on Nutrient Agar supplemented with yeast extract. All plates were incubated at 37^0C for a maximum of 48hrs.

2.8. Enumeration Of *Rhizobium* Sp. Populations In Each Soil

Congo red Yeast-extract Mannitol Agar (YMA) (Hann, 1966) was inoculated with 1g of each soil treatment type. Incubation was at 30^0C for 5 days.

2.9. Determination Of Bacterial Diversity In The Soils

The Phene Plate (PhP) System which deals with Finger Printing of Bacteria in Microplates (Kuhn *et*al, 1991; Kuhn & Mollby, 1993) was used in the determination.

The bacteria to be tested were first pre-cultivated on appropriate agar media such as Blood agar, Brain Heart Infusion agar, Brilliant Green agar, Cereus selective agar, Deoxycholate Citrate agar, Eosin Methylene Blue agar, KF Streptococcus agar, MacConkey agar, Nutrient agar, Standard Plate Count agar, SS agar, Staphylococcus Medium, Triple Sugar Iron. The same pre-cultivation conditions were used for all strains in the test series.

A multichannel pipette with sterile tips was used to fill all wells in the PhP plate with suspending substrate. Aliquots of 0.320 - 0.375ml of the substrate were dispensed into all eight wells of 'Column 1' in the plate, and 0.150ml into all the other wells. All wells in 'Column 1' were inoculated with eight different types of bacteria colonies. The plates were left for at least one hour, after which the bacterial suspensions in the first column were homogenized with the aid of the multichannel pipette. Quantities of 25μl of the bacterial suspensions in the first column were then transferred to all the other wells in each row with the multichannel pipette. Colonies suspected to be anaerobic were covered with sterile paraffin oil. Each plate was covered by a sterile lid and put in a wet chamber to avoid drying. The plates were incubated at 37^{0}C. The colour of each well was assessed after 16, 40 and 64 hours of incubation. An optical microplate reader connected to a computer with the PhP software was used. Three readings were made, after 16, 40 and 64 hours, respectively. The absorbance was measured at 620nm.

2.10. Statistical Analysis

The Statgraphics Plus for Windows version 4.0 was employed to test for significant differences between the various means of parameters of the differently treated soils and those of the untreated soil.

3. RESULTS AND DISCUSSION

3.1. Enumeration Of Population Of Rhizobia In The Differently Treated Soil Types

Total viable count studies using Congo red YMA produced the *Rhizobium* sp. population numbers indicated in Table 3. There were high population numbers per gram of soil in the cerox-treated soil and the non-treated soil, respectively. Ceresate- and paraquat-treated soils had very low population numbers per gram of soil. There was no statistically significant difference between the means of the population numbers in the ceresate-treated and paraquat-treated soils. There was, however, statistically significant difference in the means of the population numbers in the non-treated soil, cerox-treated soils and those of the ceresate-treated and paraquat-treated soils.

Table 3. Rhizobium spp. population numbers in soils with the different treatments.

Treatment Type Soil	Mean number of *Rhizobium*spp. Population (g^{-1}soil) x 10^4
Cerox-treated soil	138 (± 12.11)
Ceresate-treated soil	20 (± 4.11)
Paraquat-treated soil	12 (± 3.55)
Non-treated soil	180 (± 9.99)

3.2. Assessment Of Extent Of Growth Of Experimental Plants

Plants were assessed four weeks after germination. On the basis of plant growth and the extent of nodulation recorded in Table 4, the plants could be described as follows:

- Non-treated soil:- plants grew well with deep green foliage, had highest mean leaf number and mean leaf length. They also formed the highest number of nodules and were the largest.
- Cerox-treated soil:- plants showed moderate growth and nodulation.
- Ceresate- and paraquat-treated soils:- plants showed stunted growth and yellowish-green foliage and formed the smallest number of mean number of nodules per plant, 3 – 10 nodules, as compared to 30 – 44 mean nodules per plant of the non-treated soil and the cerox-treated soil.

There was no statistically significant difference between the means of leaf length of plants cultured in cerox-treated soil and those of plants in paraquat-treated soil. There was also no significant difference between means of leaves of plants in cerox-treated soil and those in the non-treated soil.

There was no statistically significant difference between the mean leaf numbers of plants cultured in cerox- and ceresate-treated soils. There was no significant difference between the mean leaf sizes of plants cultured in cerox-treated soil, paraquat-treated soil and the non-treated soil. There was also no significant difference between the mean leaf sizes of plants cultured in the cerox-treated soil and the ceresate-treated soil.

There was no significant difference between the mean number of nodules of plants cultured in ceresate- and paraquat-treated soils.

Table 4. Growth and nodulation of the bambara groundnuts raised in the differently treated soils.

Treatment Type	Mean leaf length (cm)	Mean leaf Number	Mean leaf size (cm)	Mean number of nodules
Cerox-treated	6.13 (±0.79)	11 (±2.94)	2.29 (±0.55)	30 (±5.16)
Ceresate-treated	3.36 (±0.86)	10 (±1.15)	1.57 (±0.28)	3 (±2.83)
Paraquat-treated	5.45 (±0.55)	6 (±2.16)	2.06 (±0.39)	8 (±5.45)
Non-treated soil	6.36 (0.40)	15 (±2.50)	2.37 (±0.24)	44 (±5.72)

3.3. Total Heterotrophic Bacteria Counts In Soils After Chemical Application

The results in Table 5 show that all the differently treated soil types had total viable bacteria present. The mean number of viable heterotrophic bacteria recorded for the soil samples varied from 40 x 10^4cfu g^{-1} to 61 x 10^5 cfu g^{-1}. Paraquat-treated soil recorded the least number of heterotrophic bacteria followed by Ceresate and the Cresox. The non-treated soil recorded the highest mean viable heterotrophic bacteria count. There was statistically significant difference between the means of the 4 variables at the 95.0% confidence level with the treatment types. There was, however, no significant difference between the various means before treatment, treatment after day 1, and treatment after 3 weeks at the 95.0% confidence level.

Table 5. Mean values of total heterotrophic bacteria count surviving after treatment with Agrochemicals.

Treatment Type	Value Before Treatment	Value after Treatment 1 Day	Value After Treatment 3 weeks
Cerox	60×10^5 (±90.65)	18×10^5 (±258.20)	10×10^5 (±182.57)
Ceresate	40×10^5 (±75.28)	73×10^4 (±29.44)	70×10^4 (±52.28)
Paraquat	56×10^5 (±45.09)	41×10^4 (±54.77)	40×10^4 (±65.83)
Non-treated	52×10^5 (±19.90)	54×10^5 (±496.66)	61×10^5 (±258.20)

3.4. Analysis Of The Diversity Indices Of The Bacterial Flora In The Soils

The diversity indices of the bacterial flora were high (more than 0.90) for both the non-treated soil and the cerox-treated soil (Tables 6a and 6b). However, the diversity indices of the bacterial flora for the ceresate- and paraquat-treated soils had values of less than 0.90 (Table 6c and 6d). A high Di (maximum value is +1) means that the assayed isolates were evenly distributed into different types, whereas low Di (minimum value is 0) means that one or few types of bacteria dominated the studied population (Kuhn *et al,* 1991; Kuhn & Mollby, 1993; Ampofo & Clerk, 2003).

3.5. Similarities Between The Bacterial Populations In The Differently Treated Soil

The PhP software used (Kuhn & Mollby, 1993) also calculated the population similarity coefficients (Sp) between the different treatments. Sp coefficients were performed according to the unweighted-pair group method using the average linkages method. High Sp coefficients (<0.5) means that the two compared samples shared many identical genera. Low Sp coefficients (>0.5) means different bacterial populations (Sneat & Sokal, 1973). The mean similarities are presented in Table 7. Comparison between the differently treated soil types, i.e. cerox-treated soil, ceresate-treated soil, and paraquat-treated soil showed Sp values all below 0.50, an indication of related populations with high diversity indices of bacteria. Comparison between the populations of bacteria from the non-treated soil and the agrochemically treated soils, however, showed Sp values greater than 0.5, an indication that the related populations were of low diversity indices Kuhn *et* al, 1991; Kuhn & Mollby 1993; Ampofo & Clerk, 2003; Gabrielson *et* al, 2003).

This study has confirmed detrimental effect of insecticide on bacterial populations in the soil. Total heterotrophic counts, rhizobial counts as well as the number of nodules of all samples taken from the chemically treated soils were all low as compared to values obtained for the untreated soil. However the effect of the insecticide was minimal in all cases as compared to the effects of the herbicides on the soil fauna.

Chemicals exert number of different toxic effects on a bacterial cell. It is difficult or even impossible to deduce the toxic mechanism of a specific chemical by just looking at its molecular structure, although chemicals with similar structures and/or physico-chemical properties are expected to have similar modes of action. Several studies have been done on the Quantitative Structure-Activity Relationships (QARs), but still knowledge is scarce. There are general rules though, such as lipophilic chemicals being more prone to disturb the bacterial membrane than hydrophilic chemicals, and electrophilic chemicals often forming irreversible covalent bonds to their target site at nucleophilc entities in biological molecules, such as proteins and DNA. A chemical may have multiple modes of toxic action and at low concentration it may even be used as a nutrient.

The effect of glyphosate on soil microbes has been studied by several authors because glyphosate, unlike most other herbicides, kills the plant by blocking a biochemical pathway which is also essential for most of the bacteria and fungi. It is known (Eberbach & Douglas, 1983) that glyphosate blocks certain biochemical pathways that are essential for growth of bacteria and the low number in bacteria population (10×10^5) as compared to the population in the non-treated soil (61×10^5) is evident enough to support this. In similar experiments conducted in Australia by the CSIRO Lands and Water the herbicides Ally®, Hoegrass® and Paraquat® were applied directly onto soil without any stubble cover at two and five times the recommended rate. In most situations this low level of functioning continued up to nine weeks. However, when the chemical was applied directly to the soil or to growing plants, the stress time for soil organisms was reduced. The research showed that it takes six weeks for the microbial activity to return to normal.

Table 6. a) Diversity among bacterial flora in the non-treated soil,b) Diversity among bacterial flora in the cerox-treated soil, c) Diversity among bacterial flora in the ceresate-treated soil, d) Diversity among bacterial flora in the paraquat-treated soil.

Sample Name and No.		No. of isolates	Di value
Non-treated soil	1	24	0.992
	2	24	0.908
	3	24	0.974
	4	24	0.962
Mean Diversity			**0.959**
		a)	

Sample Name and No.		No. of isolates	Di value
Non-treated soil	1	24	0.989
	2	24	0.978
	3	24	0.987
	4	24	0.962
Mean Diversity			**0.980**
		b)	

Sample Name and No.		No. of isolates	Di value
Non-treated soil	1	24	0.898
	2	24	0.862
	3	24	0.855
	4	24	0.915
Mean Diversity			**0.880**
		c)	

Sample Name and No.		No. of isolates	Di value
Non-treated soil	1	24	0.842
	2	24	0.814
	3	24	0.882
	4	24	0.880
Mean Diversity			**0.850**
		d)	

Table 7. Similarities between the bacterial populations for the different soil treatment types.

Parameter	Population of	Compared to	Sp value
Soil treatment type	Cerox	Ceresate	0.35
	Cerox	Paraquat	0.49
	Ceresate	Paraquat	0.13
No treatment	Non-treated	Cerox	0.51
	Non-treated	Ceresate	0.55
	Non-treated	Paraquat	0.52

The Phene-Plate (PhP) system for biochemical finger printing of bacteria, which is based on measurements of the kinetics of biochemical tests, was suitable in using to type total of 384 isolates of bacteria in microplates. The system included mathematical models and had the advantage of calculating the diversity index (Di) of the bacterial populations present in each of the treated soils, as well as calculating the similarity coefficient (Sp) (Kuhn *et* al, 1991; Kuhn & Mollby, 1993) between the populations of bacteria in the different treatments.

The current gaps in knowledge about the characteristics effects and exposure patterns of existing chemicals must be filled. Given the large knowledge gaps about chemicals on the market, it is important to generate and assess information regarding their potential risks by means of appropriate legal and regulatory instruments, voluntary agreements and economic incentives. A scientific, rules-based approach requires reliable information on effects and exposure as the basis for risk management decisions, where such information is not available, more and more countries may take precautionary approach. Workers and the public must take a more active role in monitoring and contributing to chemical safety management discussions. To facilitate this, good data from research institutions on health and environmental impacts must be more widely available.

Policies need to be established to ensure that this information is reliable, and presented in a way that is useful to all potential users for decision-making, including workers, the general public and the government. Further, governments and industry should work toward educating the public with respect to chemical safety and, where feasible, provide public interest groups with resources that allow them to play the equitable role in policy discussions.

The half-life for glyphosate is between 3 -130 days (Eberbach & Douglas, 1983), hence the effect of glyphosate on the soil bacteria was still evident. Dimethoate has a half-life of about 3 - 5 days (Bohn, 1964), but its effect on bacteria growth was still evident after three weeks. This means, dimethoate had still not been degraded or the recovery rate for the microorganisms was very slow. The half-life of paraquat is not known, but it is known that paraquat is not easily degraded and sticks to the surface soil for a longer period, because it is not leached easily. Pots treated with this agrochemical showed the effects after three weeks of application.

The following recommendations are made from this study

- Agricultural biotechnology may offer a possible alternative that may permit higher yield levels without intensive use of agrochemicals.

- If the herbs to be eliminated can easily be uprooted, then it is more advisable to do manual elimination, especially in small farming systems.

ACKNOWLEDGEMENTS

This investigation received financial support from the Government of Ghana. The authors would like to thank the Director-General of the Council for Scientific and Industrial Research of Ghana, the Director of the Water Research Institute, the Head of Environmental Biology and Health Division and the Head of Department of Botany, University of Ghana for their assistance. The cooperation and efforts of the staff of the Microbiology Section of the Water Research Institute are acknowledged.

REFERENCES

1. Aguilera-del realA.; Valverde-García, A.; Fernandez-Alba, A.R.; Camacho-Ferre, F. (1997Behaviour of endosulfan residues in peppers, cucumbers and cherry tomatoes grown in Greenhouse: Evaluation by decline curves. *Pestic. Sci.* 51194200

2. M Alexander, 1979Introduction to Soil Microbiology. Kreiger Publishing Company. 3

3. Ampofo JA & Clerk GC (2003Diversity of bacteria in sewage treatment plant used as fish culture pond in southern Ghana*Aquaculture Research*, 3419

4. G. F Antonious, M. E Byers, 1997Fate and movement of endosulfan under field conditions. *Environ. Toxicol. Chem.* 16644649

5. W. R Bohn, 1964The disappearance of dimethoate from soil. J. Econ. Entomol., 57(6), 798799

6. L. E Castillo, E De La Cruz, and C Ruepert, 1997Ecotoxicology and pesticides in tropical aquatic ecosystems of Central America. *Environ. Toxicol. Chem.* 164151

7. L. E Castillo, E Martínez, C Ruepert, C Savage, M Gilek, M Pinnock, and E Solis, 2006Water quality and macroinvertebrate community response following pesticide applications in a banana plantation, Limon., Costa Rica. *Science of the Total Environment* 367418432

8. C. M Cooper, 1993Biological effects of agriculturally derived surface-water pollutants on aquatic systems-a review. J. Environ. Qual. 22402408

9. J De Reuck, F Colardyn, J Willems, 1979Fatal encephalopathy in acute poisoning with organophosphorus insecticides. A Clinicopathologic study of two cases. Clin. Neurol. Neurosurg., 81 (4), 247254

10. M. E Delorenzo, G. I Scott, and P. E Ross, 2001Toxicity of pesticides to aquatic microorganisms: A review. *Environ. Toxicol. Chem.* 208498

11. Eberbach PL & Douglas LA (1983Persistence of glyphosate in a sandy loam *Soil Biology and Biochemistry* 15, 485487

12. L Edwards, 1975Effects of pesticides on soil organisms. *Soil Biology and Biochemistry*. 12, 285

13. J Gabrielson, I Kuhn, P Colque-navarro, M Hart, A Iversen, McKenzie D & Mollby R (2003Microplate-based microbial assay for risk assessment and (eco)toxic fingerprinting of chemicals. *Analytica Chimica Acta*, 485, 121130

14. A Gerken, J. V Suglo, and M Braun, 2001Pesticide policy in Ghana. MoFA/PPRSD, ICP Project, Pesticide Policy Project/GTZ, Accra.

15. H Goebel, S. G Gorbach, W Knauf, R. H Rimpau, H Huttenbach, 1982Properties, effects, residues, and analytics of the insecticide endosulfan.*Residue Rev.*, 831165

16. N. J Hann, 1966The Congo Red reaction in bacteria and its usefulness in the identification of rhizobia. *Canadian Journal of Microbiology*, 12, 725733

17. R Kennedy, F Sánchez-bayo, S. W Kimber, L Hugo, N Ahmad, 2001Off-site movement of endosulfan from irrigated cotton in New South Wales.*J. Environ. Qual.* 30683696

18. C Koppel, Z Forycki, K Ibe, 1986Hemoperfusion in severe Dimethoate poisoning. Intensive Care Med., 12 (2), 110112

19. Kuhn I & Mollby R (1993The PhP RS system- a simple microplate method for studying coliform bacterial populations *Journal of Microbiol. Methods* 17255259

20. I Kuhn, G Allestam, Stenstrom TA & Mollby R. (1991Biochemical fingerprinting of water coliform bacteria- a new method for measuring the phenotypic diversity and for comparing different bacterial populations. Appl. Environ. Microbiol., 57 (11), 31713177

21. H Maier-bode, 1968Properties, effect, residues and analytics of the insecticide endosulfan. *Residue Rev.* 22244

22. M Mcertensson, 1992Effects of agrochemicals and heavy metals on fast-growing rhizobia and their symbiosis with small-seeded legumes. *Soil Biology and Biochemistry*. 24, 435445

23. McEwen FL & Stephenson GR (1979) Uses and Significance of Pesticides in the EnvironmentA Wiley Interscience Publication, 229252

24. J. B Moorman, 1989A review of pesticide effects on microorganisms and microbial processes related to soil fertility. *J. Prod. Agric.*, 21, 1423

25. M Neumann, and D Dudgeon, 2002The impact of agricultural runoff on stream benthos in Hong Kong, China. *Water Res.* 3631033109

26. G. K Nurah, 1999A baseline study of vegetable production in Ghana. National Agricultural Research Project (NARP) Report, Accra, 151 pp.

27. M. L Rueppel, B. B Brightwell, J Schaefer, J. T Marvel, 1977Metabolism and degradation of glyphosate in soil and water. *Agriculture and Food Chemistry* 25, 517528

28. I Schuphan, K Ballschmiter, G Tolg, 1968Zum metabolismus des endosulfans in ratten and mausen. *Z. Naturforsch. B 23*701706

29. N Sethunathan, M Megharaj, Z Chen, N Singh, R. S Kookana, R Naidu, 2002Persistence of endosulfan and endosulfan sulfate in soil as affected by moisture regime and organic matter addition. *Bull. Environ. Contam. Toxicol.*, 68725731

30. P. K Shetty, J Mitra, N. B. K Murthy, K. K Namitha, K. N Savitha, K Raghu, 2000Biodegradation of cyclodiene insecticide endosulfan by *Mucor thermohyalospora* MTCC 1384. *Curr. Sci., 79*13811383

31. Sneath PHA & Sokal RR (1973) Numerical Taxonomy: The Principles and Practice of Numerical ClassificationW.H. Freeman, San Francisco.

32. C. M Tu, 1990Effect of four experimental insecticides on enzyme activities and level of adenosine triphosphate in mineral and organic soils. *J. Health Environ. Sci.*, B25 (6), 787800

33. U.S.Department of Health and Human Services (1990). Toxicological profile for endosulfan. Agency for toxic substance and disease registry, Atlanta.

34. M Waiwright, 1978A review of the effect of pesticides on microbial activity in soil. *J. Soil. Sci.*, 29, 287298

35. J Zulalian, 1990Study of the absorption, excretion, metabolism, and residues in tissues of rats treated with carbon-14-labeled pendimethalin, Prowl herbicide. *J. Agric. Food Chem.*, 38, 17431754

Chapter 4

Levels of essential and non-essential metals in ginger (*Zingiber officinale*) cultivated in Ethiopia

Yohannes Wagesho and Bhagwan Singh Chandravanshi*

Department of Chemistry, College of Natural Sciences, Addis Ababa University, P.O. Box 1176, Addis Ababa, Ethiopia

ABSTRACT

Background

Ginger (*Zingiber officinale* Roscoe) is a common condiment for various foods and beverages and widely used worldwide as a spice. Its extracts are used extensively in the food, beverage, and confectionary industries in the production of products such as marmalade, pickles, chutney, ginger beer, ginger wine, liquors, biscuits, and other bakery products. In Ethiopia, it is among the important spices used in every kitchen to flavor stew, tea, bread and local alcoholic drinks. It is also chiefly used medicinally for indigestion, stomachache, malaria, fevers, common cold, and motion sickness. The literature survey revealed that there is no study conducted on the determination of metals in ginger cultivated in Ethiopia. Hence it is worthwhile to determine the levels of essential and non-essential metals in ginger cultivated in Ethiopia.

Methods

The levels of essential (Ca, Mg, Fe, Zn, Cu, Co, Cr, Mn, and Ni) and non-essential (Cd and Pb) metals in ginger (*Zingiber officinale* Roscoe) cultivated in four different regions of Ethiopia and the soil where it was grown were determined by flame atomic absorption spectrometry. 0.5 g of oven dried ginger and soil samples were digested using 3 mL of HNO_3 and 1 mL of $HClO_4$ at 210°C for 3 h and a mixture of 6 mL aqua-regia and 1.5 mL H_2O_2 at 270°C for 3 h, respectively.

Results

The mean metal concentration (μg/g dry weight basis) ranged in the ginger and soil samples, respectively, were: Ca (2000–2540, 1770–3580), Mg (2700–4090, 1460–2440), Fe (41.8–89.0, 21700–46900), Zn (38.5–55.2, 255–412), Cu (1.1–4.8, 3.80–33.9), Co (2.0–7.6, 48.5–159), Cr (6.0–10.8, 110–163), Mn (184–401, 1760–6470), Ni (5.6–8.4, 14.1–79.3) and Cd (0.38–0.97, 0.24–1.1). The toxic metal Pb was not detected in both the ginger and soil samples.

Conclusion

There was good correlation between some metals in ginger and soil samples while poor correlation between other metals (Fe, Ni, Cu). This study revealed that Ethiopian gingers are good source of essential metals and free from toxic metal Pb while containing negligible amount of Cd.

KEYWORDS

Ginger, *Zingiber officinale*, Essential metals, Non-essential metals, Soil Ethiopia

1. INTRODUCTION

The delightful flavour and pungency of spices make them indispensable in the food industry to flavour, improve and increase the appeal of their products. Spices impart aroma, colour and taste to food preparations and sometimes mask undesirable odours. In addition, they are reputed to possess several medicinal and pharmacological properties and hence find position in the preparation of a number of medicines (Parthasarathy et al. 2008).

Ethiopia is among the largest consumer of spices in Africa. The major use of spices in the country is in the preparation of a highly spiced stew known as 'Wot' which together with 'Injera' is consumed by a large proportion of the population everyday as their main food. In addition, spices are also used by the numerous ethnic groups in the country to flavor bread, meat, soups, different vegetables, and as medicines and perfumes (Asfaw and Demissew 2009).

The spice ginger obtained from the underground stems or the rhizome of *Zingiber officinale* Roscoe, one of the most widely used species of the family Zingiberaceae, is a common condiment for various foods and beverages. Both fresh and dried ginger rhizomes are used worldwide as a spice, and ginger extracts are used extensively in the food, beverage, and confectionary industries in the production of products such as marmalade, pickles, chutney, ginger beer, ginger wine, liquors, biscuits, and other bakery products (Mishra 2009). Ginger is also widely used in both traditional and contemporary natural medicine (Zingiber officinale 2010). In Ethiopia, it is among the important spices used in every kitchen to flavor stew, tea, bread and local alcoholic drinks (Asfaw and Abegaz, 1995). It is also chiefly used medicinally for indigestion, stomachache, malaria, fevers, common cold, and motion sickness.

The major ginger growing area in Ethiopia includes wetter regions at altitude below 2000 m in Kefa, Illubabur, Gamo Gofa, Sidama, Wellega, Wolaita, and Kembata-Tambaro. Currently, it has become an important cash crop for farmers in southern and south-western parts of Ethiopia. The production of this spice has been expanding in most parts of the country, as it can be grown under varied climatic conditions. It thrives well in areas with altitudes from sea level to 1500 m, mean annual temperature of 20–32°C and with total rainfall greater than 1200 mm. Well-drained, fertile and friable soil with enough humus and neutral pH is the ideal soil type for the production of ginger (Asfaw and Demissew 2009; Hailemicheal et al. 2008).

The unique flavor properties of ginger arise from the combination of pungency and aromatic essential oil. The main pungent compounds in fresh ginger are a series of homologous phenolic ketones known as gingerols. The gingerols are thermally unstable and are converted under high temperature to shogaol (Mishra 2009). Shogaols, which are more pungent than gingerols, are the major pungent compounds in dried ginger (Jolad et al. 2004).

The ginger rhizome also contains resin, proteins, cellulose, pentosans, starch and mineral elements. Of these, starch is the most abundant and comprises 40–60% of the rhizome on a dry weight basis. The relative abundance of certain constituents of ginger is determined by the cultivar grown, the environmental conditions of growth and the stage of maturity at harvest (Parthasarathy et al. 2008).

Humans require a suite of mineral elements in varying amounts for proper growth, health maintenance and general well being. Minerals are essential components of our diet that serve as cofactors in the thousands of enzyme-controlled reactions, control the action of nerves and muscles, help maintain the body's water balance, and buffer the pH (acidity) of the cell and extracellular fluids (Minerals-Learn 2010). Plant-derived foods have the potential to serve as dietary sources for all human-essential minerals that make a significant contribution to daily mineral needs at all stages of the life cycle (Grusak 2002). Generally, too low or too high of a concentration of trace elements in human diet can affect the quality of human life.

Mineral uptake by plants can be affected by several factors including mineral concentrations in soils, soil pH, cation exchange capacity, organic matter content, types and varieties of plants, and age of the plant (Jung 2008). In ideal word we would take in our daily requirement of minerals by eating plants that grow in mineral rich soils (Harold and Leslie 2000; Lokeshwari and Chandrappa 2006). The contamination of soil by atmospheric deposition of toxic metals affects soil properties and further increases plant metal levels through root uptake (Pandey and Pandey 2009), and eventually these metals are taken up by plants parts and transfer some into the food chain. Consequently, higher soil heavy metal concentration can result in higher levels of uptake by plants (Ebong et al. 2008).

Contamination and subsequent pollution of the environment by toxic heavy metals have become an issue of global concern due to their sources, widespread distribution and multiple effects on the ecosystem. Heavy metals are generally present in agricultural soils at low levels. Due to their cumulative behaviour and toxicity, however, they have a potential hazardous effect not only on crop plants but also on human health (Uwah et al. 2009). Therefore, a comprehensive study related to the assessment of levels of essential and toxic metals of plants and soil where the plant has grown is crucial with respect to human health and the quality of its products.

Some studies have been carried out on levels of essential and non-essential metals in some spices and medicinal plants cultivated in Ethiopia (Derbie and Chandravanshi 2011; Gebre and Chandravanshi 2012; Kitata and Chandravanshi 2012; Weldegebriel et al. 2012; Aregahegn et al. 2013; Mekebo and Chandravanshi 2014; Endalamaw and Chandravanshi 2015; Dubale et al. 2015). Some studies have also been conducted on determination of essential and non-essential metal levels of ginger in Nigeria (Obiajunwa et al. 2002; Ogunwandea and Olawore 2004; Aiwonegbe and Ikhuoria 2007), India (Devi et al. 2008), and Saudi Arabia (Al-Eed et al. 2002; Alwakeel 2008). However, the literature survey revealed that there is no study conducted on the determination of metals in ginger cultivated in Ethiopia. Since ginger is used as a spice for many peoples of Ethiopia and it is cash crop, the knowledge of their mineral levels is of particular interest. Hence it is worthwhile to determine the levels of essential and non-essential metals in ginger cultivated in Ethiopia. Therefore, this study deals with assessment of levels of metals (essential and non-essential) in ginger cultivated in Ethiopia and it aims to fill the gap at least partially in the area and initiate others for further study on ginger and closely related plants widely used throughout the country. The outcome of this study will ultimately help to ensure the dietary safety of the society and improving the country's economy by increasing both quality and quantity of ginger cultivated in Ethiopia.

The aim of this study was to determine the levels of selected major, trace and toxic metals (Ca, Mg, Fe, Zn, Cu, Co, Cr, Mn, Ni, Cd, and Pb) in ginger cultivated in Ethiopia, to assess the level of minerals in soil samples where the ginger was grown and to correlate the levels of minerals in the ginger with that of soil in which it was cultivated. It was also aimed to compare the levels of metals in the ginger cultivated in Ethiopia with the levels of metals in the ginger from other countries.

2. METHODS

2.1 Equipments and reagents

Chopping board (PTFE, China) and Teflon (PTFE) knife were used to cut ginger rhizome in to pieces. A drying oven (Digitheat, J.P. Selecta, Spain) was used to dry ginger sample. Electronic blending device (Moulinex, France) was used for grinding and homogenizing the sample to determine the total metal content of the ginger. Mortar and pestle was used to grind soil sample. Analytical balance (Larko, LA114, 110 g/0.1 g) with precision of ± 0.0001 g was used to weigh the ginger and soil sample. A 100 mL round bottomed flasks fitted with reflux condensers were used in Kjeldahl apparatus hot plate to digest the dried and powdered ginger and soil samples. Borosilicate volumetric flasks (25, 50 and 100 mL) were used during dilution of sample and preparation of metal standard solutions. Measuring cylinders (Duran, Germany), pipettes (Pyrex, USA), and micropipettes (Dragonmed, 1–10 µL, 100–1000 µL, Shangai, China) were used during measuring different quantities of volumes of sample solution, acid reagents and metal standard solutions. Flame atomic absorption spectrophotometer, FAAS (Buck Scientific Model 210VGP AAS, East Norwalk, USA) equipped with deuterium arc back ground correctors and hollow cathode lamps with air-acetylene flame was used for the determination of the metals (Ca, Mg, Cu, Zn, Mn, Ni, Fe, Co, Cr, Pb and Cd) in the ginger and soil samples.

All the reagents used were of analytical grade. HNO_3 (69–72%) and $HClO_4$ (70%) (Research Lab Fine Chem Industries, Mumbai, India) were used for the digestion of ginger samples. Aqua-regia prepared from 3:1 ratio of 37% HCl (Riedel-deHaën, Germany) and (69–72%) HNO_3, and extra pure hydrogen peroxide 30% H_2O_2, (Scharlau, European Union), were used for the digestion of soil sample. Lanthanum nitrate hydrate (98%, Aldrich, USA) was used to avoid refractory interference (for releasing calcium and magnesium from their phosphates). Stock standard solutions containing 1000 mg/L, in 2% HNO_3, of the metals Ca, Mg, Cu, Zn, Mn, Ni, Fe, Co, Cr, Pb and Cd (Buck Scientific Puro-Graphic, USA) were used for preparation of calibration standards and in the spiking experiments. Deionized water (chemically pure with conductivity 1.5 µs/cm and below) was used for dilution of sample and intermediate metal standard solutions prior to analysis and rinsing glassware and sample bottles.

2.2 Sampling

Samples were collected from four major ginger producing areas in Ethiopia namely Tepi, Bombae, Hadaro and Illubabur. Tepi is located in the Sheka zone, Southern Nations, Nationalities and Peoples' Region (SNNPR), 596 km southwest of Addis Ababa with an elevation of 1097 m above sea level and lies between 7°12′–7°89′N and 35°24′ to 37°90′E. The annual mean temperature of the area ranges between 15.1–27.5°C and the annual mean rainfall ranges 1201–1800 mm. Bombae is located in Wolayita Zone, SNNPRS, 390 km south west of Addis Ababa, lies between 6°51′ and 7°35′N and 37°46′–38°1′E. Hadaro is located in Kembata zone of SPNNR. It is 270 km far from the capital city of Ethiopia which is Addis Ababa. Illubabor is located in the Oromia Region. It has a latitude and longitude of 8°18′N and 35°35′E and an altitude of 1605 m.

Fresh ginger samples were collected from the farmlands of four areas in southern and south western Ethiopia particularly Illubabur (Oromia region), Tepi (Sheka, SNNPR), Hadaro (Kambata, SNNPR), and Bombae (Wolayita, SNNPR). These sampling sites were selected based on large-scale production area of ginger in the country so that the sample partly represents the whole ginger cultivated in Ethiopia. To collect the representative sample from each sampling sites, three sub-samples (500 g each) were taken from farmlands which were roughly two km away from each other. These farmlands were randomly chosen from the three triangular corners of the area. Half kilogram of fresh ginger samples were collected from each farmland and put in clean polyethylene plastic bags labelled and brought to the laboratory for further pre-treatment. The three sub-samples were mixed together after grinding in a blender to homogenize and form bulk samples that represent each sampling areas. Finally four ginger bulk samples one from each stated areas were prepared for analysis. 12 samples with 0.5 g aliquot (three from each bulk sample) were taken for final digestion.

For comparative analysis of mineral levels in ginger and the soil where the ginger was cultivated, the soil samples were collected from the surface horizon (15–20 cm) depth of the same four sampling areas of ginger. Sampling was done similar to ginger. Half-kilogram soil samples were collected from each farmland. The three sub-samples were mixed together after grinding using a mortar and pestle to homogenize and form bulk samples that represent each sampling areas. Finally four soil bulk samples one from each stated areas were prepared for analysis. 12 samples with 0.5 g aliquot (three from each bulk sample) were taken for final digestion.

2.3 Sample preparation

Fresh ginger collected from the sampling areas was kept in plastic bags and transported to the laboratory. The rhizomes were washed with a running tap water so as to remove adsorbed soil particulates and then rinsed with deionized water. The thin outer skin of ginger was removed by plastic knife and chopped in to pieces nearly uniform size to facilitate drying uniformly. The sample was exposed to sun light for two days to reduce the moisture content. Then, to have constant mass, the sample was oven dried at 80°C for 24 h so as to express the result in terms of dry mass basis. The dried ginger was powdered using electronic blender and sieved to prepare fine powder of ginger for digestion.

The soil sample collected from the four sampling area were air dried to constant weight for three days and sieved through a 2-mm polyethylene sieve to remove large debris, stones, and pebbles. Then, the samples were ground using a mortar and pestle to pass a 500-µm sieve, homogenized, and ready for digestion.

2.4 Moisture content of ginger

To determine the moisture content of fresh ginger, first it was carefully washed with tap water to remove adsorbed soil particulates and exposed to air to vaporize water on the surface of it. Then it was weighed with electronic balance to record the initial weight with its moisture content. After oven drying at 80°C for two days it was re-weighed and re-dried until it gave constant mass. The moisture content of four samples was between 75.0–84.9%. Therefore, fresh ginger cultivated in Ethiopia has comparable moisture content with the value reported by Govindarajan (1982), which is 80.9% for typical analysis of market sample of ginger.

2.5 Digestion of ginger and soil samples

The basic requirements for sample preparation for analysis are to get an optimum condition for digestion. The optimum condition is the one which required minimum reagent volume consumption, minimum reflux time, clarity of digests, and ease of simplicity (Huang et al. 2004; Wilson et al. 2005; Demirel et al. 2008; Chen and Ma 2001; Gaudino et al. 2007; Endalamaw and Chandravanshi 2015).

In this study, to prepare a clear colorless sample solution that is suitable for the analysis using FAAS, different ginger digestion procedures were optimized using the HNO_3 and $HClO_4$ acid mixtures by varying parameters such as volume of the acid mixture, digestion time and digestion temperature (Huang et al.2004; Wilson et al. 2005; Demirel et al. 2008). The results are given in Table 1. From the optimization procedures, the acid mixture of 3 mL of HNO_3 (69–70%) and 1 mL of $HClO_4$ (70%), digestion time of 3 h and digestion temperature of 210°C were found the optimal condition for 0.5 g ginger sample. These optimum conditions were selected based on clarity of digests, minimum reagent volume consumption, minimum digestion time, simplicity and minimum temperature applied for complete digestion of sample.

A modified aqua-regia ($HNO_3 + HCl + H_2O_2$) was selected as digestion reagent for soil sample digestion in this work (Chen and Ma 2001; Wilson et al. 2005; Gaudino et al. 2007). The optimum conditions (Table 2) for soil sample digestion were a reagent mixture of 6 mL aqua-regia (3:1 ratio of HCl to HNO_3) and 1.5 mL H_2O_2, digestion temperature of 270°C and digestion time of 3 h for 0.5 g soil sample.

Applying the optimized condition, 0.5 g of dried and homogenized ginger samples were transferred into a 100 mL round bottomed flask. Then 4 mL of a mixture of HNO_3 (69–72%) and $HClO_4$ (70%) with a volume ratio of 3:1 (v/v) was added and the mixture was digested on a Kjeldahl digestion apparatus fitting the flask to a reflux condenser by setting the temperature first at 120°C for 30 min and then increased to 210°C for the remaining 2 h and 30 min. The digest was allowed to cool to room temperature for 10 min without dismantling the condenser from the flask and for 10 min after removing the condenser. To the cooled solution 15 mL of deionized water was added to dissolve the precipitate formed on cooling and to minimize dissolution of filter paper by the digest residue while filtering with Whatman, (110 mm, diameter), filter paper into 50 mL volumetric flask. The round bottom flask was rinsed subsequently with 5 mL deionized water until the total volume reached around 45 mL. To this final solution, 3 mL lanthanum nitrate solution (1% w/w) was added and the solution was filled to the mark (50 mL) with deionized water. The digestion was carried out in triplicate for each bulk sample. Digestion of a reagent blank was also performed in parallel with the ginger samples keeping all digestion parameters the same. The digested samples were kept in the refrigerator, until the levels of all the metals in the sample solutions were determined by FAAS.

Table 1. Different conditions tested for optimization of digestion procedure for 0.5 g ginger samples

Trial no.	Reagent(s) used	Reagent volume (mL)	Temp. (°C)	Digestion time (h)	Results
Optimization for reagent volume					
1	HNO_3:$HClO_4$	3:3	270	3:00	Deep yellow
2	HNO_3:$HClO_4$	4:2	270	3:00	Yellow
3	HNO_3:$HClO_4$	3:2	270	3:00	Clear yellow
4	HNO_3:$HClO_4$	4:1	270	3:00	Almost clear
5	HNO_3:$HClO_4$	2:2	270	3:00	Clear light yellow
6	**HNO_3:$HClO_4$**	**3:1**	**270**	**3:00**	**Clear and colourless**
Optimization for temperature					
1	HNO_3:$HClO_4$	3:1	150	3:00	Deep yellow
2	HNO_3:$HClO_4$	3:1	180	3:00	Light yellow
3	**HNO_3:$HClO_4$**	**3:1**	**210**	**3:00**	**Clear and colourless**
4	HNO_3:$HClO_4$	3:1	240	3:00	Clear and colourless
5	HNO_3:$HClO_4$	3:1	270	3:00	Clear and colourless
6	HNO_3:$HClO_4$	3:1	300	3:00	Clear and colourless
Optimization for digestion time					
1	HNO_3:$HClO_4$	3:1	210	1:45	Deep yellow
2	HNO_3:$HClO_4$	3:1	210	2:00	Light yellow
3	HNO_3:$HClO_4$	3:1	210	2:15	Light yellow
4	HNO_3:$HClO_4$	3:1	210	2:30	Clear light yellow
5	HNO_3:$HClO_4$	3:1	210	2:45	Clear and colourless
6	**HNO_3:$HClO_4$**	**3:1**	**210**	**3:00**	**Clear and colourless**

Rows with bold font indicate the optimal condition for the given parameter.

Table 2. Different conditions tested for optimization of digestion procedure for 0.5 g soil samples

Trial no.	Reagent(s) used	Reagent volume (mL)	Temp. (°C)	Digestion time (h)	Results
Optimization for reagent volume					
1	Aqua-regia:H_2O_2	6.5:1	300	3:00	Deep yellow with suspension
2	**Aqua-regia:H_2O_2**	**6:1.5**	**300**	**3:00**	**Light yellow with no suspension**
3	Aqua-regia:H_2O_2	5.5:2	300	3:00	Light yellow with no suspension
Optimization for temperature					
1	Aqua-regia:H_2O_2	6:1.5	240	3:00	Deep yellow with suspension
2	**Aqua-regia:H_2O_2**	**6:1.5**	**270**	**3:00**	**Light yellow with no suspension**
3	Aqua-regia:H_2O_2	6:1.5	300	3:00	Light yellow with no suspension
Optimization for digestion time					
1	Aqua-regia:H_2O_2	6:1.5	270	2:00	Deep Yellow with suspension
2	Aqua-regia:H_2O_2	6:1.5	270	2:30	Light yellow with suspension
3	**Aqua-regia:H_2O_2**	**6:1.5**	**270**	**3:00**	**Light yellow with no suspension**

Rows with bold font indicate the optimal condition for the given parameter.

Applying the optimized condition, 0.5 g of dried and homogenized soil samples were transferred into a 100 mL round bottomed flask. To this 6 mL of aqua-regia (3:1 ratio of 37% HCl to (69–72%) HNO_3, respectively) and followed by 1.5 mL of 30% H_2O_2 were added and the mixture was digested on a Kjeldahl digestion apparatus fitting the flask to a reflux condenser by setting the temperature first at 180°C for the first 30 min and then raised to 240°C for the next 30 min and finally raised to 270°C for the remaining 2 h. The rest steps were similar for both ginger and soil sample digestion procedure.

2.6 Instrument calibration

Calibration metal standard solutions were prepared for each of the metals from an intermediate standard solution containing 10 mg/L which was prepared from the atomic absorption spectroscopy standard stock solutions that contained 1000 mg/L. These intermediate standards were diluted with deionized water to obtain four working standards for each metal of interest. Then, Ca, Mg, Cu, Zn, Mn, Ni, Fe, Co, Cr, Pb and Cd were analyzed with FAAS (Buck Scientific Model 210GP) equipped with deuterium arc background corrector and standard air-acetylene flame system using external calibration curve. Three replicate determinations were carried out on each sample. All the eleven metals were determined by absorption/concentration mode and the instrument readout was recorded for each solution manually. The same analytical procedure was employed for the determination of elements in digested blank solutions.

The instrument was calibrated using four series of working standards. The working standard solutions of each metal were prepared freshly by diluting the intermediated standard solutions (10 mg/L). The correlation coefficients of the calibration curves were > 0.9999 which confirmed an excellent correlation between the absorbance and the concentration. The method detection limits were in the range 1–5 µg/g dry weight for the ginger and soil samples which indicated that the FAAS method used are applicable to detect the presence of metals in the ginger and soil samples at trace levels.

2.7 Method performance and method validation

The criteria used for evaluating analytical methods are called figures of merit. Based on these characteristics, one can predict whether a method meets the needs of intended purpose. These figures of merit are accuracy, precision, sensitivity, detection limits, and the quantitation limits (Mitra 2003).

In this study, the precision of the results were evaluated by the pooled standard deviation, and relative standard deviation of the results of nine measurements for a given bulk sample (i.e. three samples (n = 3) and triplicate readings for each sample).

In the present study, method detection limit for each metal was estimated by digesting six analytical blanks with the optimized procedure for both ginger and soil samples. Triplicate analyses of six blank samples for all elements were performed and the pooled standard deviation of the six blank reagents was calculated. The detection limits were obtained by multiplying the pooled standard deviation of the reagent blank (S_{blank}) by three ($MDL = 3 \times S_{blank}$, n = 6). The method detection limit of each metal was ≤ 5 µg/g which indicated that the method was applicable to the determination of metals at trace levels in both the ginger and soils samples.

In this work, the method validation was established by spiking experiments. The spiked samples were prepared by adding a small known quantity of metal standard solutions. For spiking ginger sample, 200 µL of 1000 mg/L Ca, 300 µL of 1000 mg/L Mg, 12.5 µL of 1000 mg/L Zn, 75 µL of 1000 mg/L Mn, 31.5 µL of 40 mg/L Ni, 10 µL of 1000 mg/L Fe, 43.5 µL of 40 mg/L Co, 25 µL of 40 mg/L Cu, 62.5 µL of 40 mg/L Cr and 30 µL of 10 mg/L Cd standard solutions were added to round bottomed flask (100 mL) containing 0.5 g ginger sample. For soil sample spiking, 350 µL of 1000 mg/L Ca, 250 µL of 1000 mg/L Mg, 235 µL of 1000 mg/L Fe, 65 µL of 1000 mg/L Mn, 75 µL of 1000 mg/L Zn, 20 µL of 1000 mg/L Ni, 200 µL of 40 mg/L Cu, 40 µL of 1000 mg/L Co, 35 µL of 1000 mg/L Cr and 32 µL of 10 mg/L Cd standard solutions were added to round bottomed flask (100 mL) containing 0.5 g soil sample. The spiked and non-spiked samples were digested and analysed in similar condition.

The results of percentage recoveries for the studied metal nutrients in both ginger and soil samples were within the acceptable range (93–106%) in the ginger and (93–107%) in the soil samples. These results verified that the optimized digestion procedure was valid for both ginger and soil sample analysis.

2.8 Statistical analysis

The analysis of variance for the equality of means and correlation between the elements in ginger and soil samples were done using the SPSS 16.0.

3. RESULTS

The concentration of eleven elements (Ca, Mg, Fe, Zn, Cu, Co, Cr, Mn, Ni, Cd, and Pb) in the digested samples of ginger and soil were determined by FAAS. Among the determined metals Pb was below the method detection limit (0.002 mg/g in the ginger and 0.004 mg/g in the soil) and the concentration of the rest of metals are shown with their respective standard deviation in Tables 3 and 4. The most abundant metal among the macro-elements in ginger was Mg followed by Ca whereas Mn content was the predominant among the tested micronutrient heavy metals followed by Fe, Zn, Co and Cu. In soil sample the most abundant metal was Fe followed by Mn, Ca, Mg, Zn, Cr, Co, Cu, Ni, and Cd.

4. DISCUSSION

4.1 Distribution pattern of the metals in ginger samples

Mineral uptake in plants is a function of mineral concentrations in soils, soil pH, cation exchange capacity, organic matter content, types and varieties of plants, and age of the plant (Jung 2008). As it can be seen from Table 3, there is large difference in concentration of different metals within ginger sample and slight variation in metals of the same type along with geographical location. The range of concentration (in mg/kg) pattern of elements in ginger sample collected from four sampling area is given in Table 3.

Table 3. Average concentration (mean ± SD, n = 3, μg/g dry weight basis) of major, trace and toxic metals in ginger samples from Tepi, Bombae, Hadaro and Illubabur

Metal	Average concentration (mean ± SD, n = 3, μg/g dry weight basis)			
	Tepi ginger	Bombae ginger	Hadaro ginger	Illubabur ginger
Ca	2000 ± 47	2540 ± 93	2190 ± 24	2490 ± 41
Mg	2990 ± 9	2700 ± 57	2760 ± 11	4090 ± 105
Cu	4.78 ± 0.34	1.86 ± 0.18	2.53 ± 0.19	1.10 ± 0.05
Zn	55.2 ± 3.9	39.6 ± 0.5	38.5 ± 0.5	54.0 ± 2.7
Mn	385 ± 9	285 ± 4.3	184 ± 3.6	401 ± 12
Ni	5.61 ± 0.44	5.46 ± 0.48	6.78 ± 0.53	8.40 ± 0.32
Fe	44.2 ± 3.3	55.4 ± 5.0	41.8 ± 2.8	89.0 ± 6.1
Co	7.58 ± 0.46	5.68 ± 0.40	2.04 ± 0.14	2.18 ± 0.18
Cr	9.28 ± 0.61	6.02 ± 0.14	9.17 ± 0.62	10.8 ± 0.2
Cd	0.97 ± 0.08	0.38 ± 0.02	0.38 ± 0.02	0.70 ± 0.07
Pb	ND	ND	ND	ND

ND: Concentration of the tested heavy metal was below the method detection limit.

Table 4. Average concentration (mean ± SD, n = 3, µg/g dry weight basis) of major, trace and toxic elements in soil samples from Tepi, Bombe, Hadaro and Illubabur

Metal	Average concentration (mean ± SD, n = 3, µg/g dry weight basis)			
	Tepi soil	Bombae soil	Hadaro soil	Illubabur
Ca	3580 ± 16	2060 ± 10	2040 ± 43	1770 ± 39
Mg	2430 ± 141	1460 ± 45	1660 ± 8	2440 ± 8
Cu	33.9 ± 0.5	3.76 ± 0.07	6.77 ± 0.17	33.7 ± 0.8
Zn	389 ± 36	344 ± 28	255 ± 14	413 ± 39
Mn	6470 ± 81	1760 ± 26	1920 ± 28	4680 ± 32
Ni	79.3 ± 1.2	14.1 ± 0.3	21.4 ± 1.0	73.1 ± 4.7
Fe	46900 ± 600	21800 ± 821	21700 ± 407	46170 ± 484
Co	159 ± 2.8	57.1 ± 2.1	48.5 ± 0.6	132 ± 1.9
Cr	139 ± 12	110 ± 7.6	114 ± 1.8	163 ± 2.5
Cd	1.08 ± 0.08	0.24 ± 0.02	0.73 ± 0.06	1.20 ± 0.07
Pb	ND	ND	ND	ND

ND: Concentration of the tested heavy metal was below the method detection limit.

Ginger contained higher amount of Mg (2700–4090 mg/kg), followed by Ca (2000–2540 mg/kg). The higher levels of Mg in the ginger is probably due to the fact that nutrient elements such as N, P, K, S, and Mg are highly mobile in the plant tissue and trans-located from old plant tissue to new plant tissue. The other probable reason for higher concentration of Mg and Ca is if the soil which have been used for cultivating the plant, are highly fertilized with manure and organic residues, K, Ca and Mg are highly available for plant uptake. Hence, the plant has high amount of these metals.

Mn (184–401 mg/kg) was the most accumulated trace metal followed by Fe (41.8–89.0 mg/kg) and Zn (38.5–55.2 mg/kg) in ginger sample. Higher Mn levels in the ginger may be attributed to the availability of this micronutrient heavy metal in relatively acidic soils of the farmland. The chemical forms of Mn present in soil are known to depend on soil pH. In acidic soil, the easily absorbed form, Mn^{+2} released from soil by H^{+}, which is produced from $NH_4{}^{+}$ (Zubillaga et al. 2008), can be readily taken up and accumulated in the ginger.

It has been reported that Fe and Zn are the main elements that plant could accumulate and pass up in the food chain. Therefore, the high concentration of Zn from trace metals next to Mn and Fe in ginger may be because of the fact that these ions are readily transferred from the soil to plants, and accumulated in plants.

The levels of other essential trace metals detected in ginger were Co (2.04–7.58 mg/kg), Ni (5.61–8.40 mg/kg), Cr (6.02–10.8 mg/kg) and non-essential heavy metals Cd (0.38–0.97 mg/kg). The level of Cd was the least among the metals; however due to its toxicity deserves special concern. The non-essential heavy metal, Pb, was found to be below the method detection limit. In general the concentration pattern of metals in ginger was decreased as Mg > Ca > Mn > Fe > Zn > Cr > Ni > Co > Cu > Cd.

As it can be seen from Table 3, ginger are a good source of major and trace metals that are essential to human in addition to its food flavouring purpose. The small amount of Co and Cu found in ginger does not contradict with the requirement of the metal for proper functioning of the body, because these metals are required in small amount (Co = 0.3 mg/day as a constituent of vitamin B_{12} and Cu = 3.5 mg/day).

4.2 Daily intake of metals from ginger

Daily intake of metals from ginger has been calculated based on the assumption that an adult person consumes 5 g fresh ginger per day in different form. The moisture content of Ethiopian fresh ginger was found in the range 75–85% with a mean value of 80%. Thus the dry mass of 5 g ginger is 1 g. The metal contents of 1 g dry ginger are given in Table 5. For comparison the recommended or adequate daily intake and the allowable upper limit of daily intake of metals are also given in the Table 5. The data in the table clearly show that the daily intake of Ca, Mg and Cu are well below the recommended daily intake while Zn, Mn, Fe, and Cr exceed the recommended daily intake but are well below the allowable upper limit of daily intake of these metals. Daily intake of Ni and Co from Ethiopian ginger is very small and well below the allowable daily intake. Daily intake of toxic metal Cd from Ethiopian ginger is negligible and Pb is none (not present in Ethiopian ginger). Therefore, it can be concluded that Ethiopian ginger is good source of essential metals and free from toxic metal and hence safe for human consumption.

4.3 Distribution pattern of metals in soil sample

The soil sample collected from four sampling areas were found to contain detectable metal content of Ca, Mg, Fe, Zn, Cu, Co, Cr, Mn, Ni, and Cd in all the four soil samples and their values are given in Table 4. Among the determined metals, Pb was found to be below the detection limit of the method used in this study. There is significant difference in concentration of different metals within soil sample and appreciable difference in the same metals of different samples. The determined concentration ranges of metals from four soil sampling area are given in Table 4.

As shown in Table 4, the concentration of Fe (21700–46900 mg/kg) in soil exceeds much the concentration of macro-elements, Ca (1770–3580 mg/kg) and Mg (1460–2440 mg/kg), this is due to the presence of excess amount of hematite (Fe_2O_3) in the soil.

Table 5. Comparison of daily intake of metals from ginger with recommended daily intake and tolerable upper limit of daily intake of metals (NebGuide **2015**)

Metal	Daily intake from ginger (mg/day)	Recommended daily intake (mg/day)	Allowable upper limit (mg/day)
Ca	2.36	1000	2500
Mg	2.00	400	350
Cu	0.0195	2	10
Zn	0.350	0.150	40
Mn	3.71	2	11
Ni	0.047	NA	1
Fe	34.1	18	45
Co	0.0992	NA	NA
Cr	0.132	0.120	NE
Cd	0.000613	None	0.0714
Pb	ND	None	None

NA = data not available. NE = not established. ND = not detected.

Concentration of Mn (1760–6470 mg/kg) an essential trace metal in these soils was higher when compared to the micronutrient heavy metals Zn (255–413 mg/kg), Cu (3.76–33.9 mg/kg), Cr (110–163 mg/kg), Co (48.5–159 mg/kg) and Ni (14.1–79.3 mg/kg). On the other hand, level of the toxic heavy metal Cd ranged from 0.24–1.08 mg/kg. The level of Pb, the other tested toxic metal, in soils of all the samples was found to be below the detection limit of the method used in this study. In general, the concentration pattern of metals in soil was decreased as Fe >> Mn > Ca > Mg > Zn > Cr > Co > Ni > Cu > Cd.

4.4 Comparisons of metal levels between ginger and soil sample

Plants absorb whatever is present in the soil medium and therefore the metals are also absorbed and become bio-accumulated in the roots, stems, fruits, grains and leaves of the plant, which may finally be transferred to man in the food chain. The sorption processes of metals by plants is significantly affected by metal level in the soil, soil pH, the presence of competing ligands, the ionic strength of the soil solution and the simultaneous presence of competing metals (Zubillaga et al. 2008).

In this work, comparative study has been established to correlate the metal level of ginger with the soil where it has grown. As it can be seen from Tables 3 and 4, for most elements (Mg, Mn, Zn, Fe (except in Tepi sample), Cu (except in Illubabur sample), Cr, and Cd) the metal levels of ginger was directly proportional to the metal levels of soil where it has grown. Taking the level of Mg as an example in both ginger and soil samples, the Mg level of ginger sample from Illubabur > Tepi > Hadaro > Bombae and the same order is true for Mg level of respective soil sample. This relation partly verifies that the metal content of the plant is a function of the metal level in the soil where it has grown. For the rest three metals (Ca, Ni and Co), the metal levels in some sampling area of ginger were varies non-proportional to levels of metal in the corresponding soil. This non-proportional variation in level of metals in ginger and soil may be resulted from the difference in availability of absorbable form of metals in soil due to difference in soil acidity or the presence of competing ligands. The accumulation factors of the metals from soil to ginger are given in Table 6.

Table 6. Accumulation coefficient of metals from the soil to ginger

Metal	Sampling sites			
	Tepi	Bombae	Hadaro	Illubabur
Ca	0.56	1.23	1.07	1.41
Mg	1.23	1.85	1.66	1.68
Cu	0.14	0.49	0.37	0.032
Zn	0.14	0.12	0.15	0.13
Mn	0.060	0.16	0.096	0.086
Ni	0.071	0.40	0.32	0.11
Fe	0.00094	0.0025	0.0019	0.0019
Co	0.048	0.099	0.042	0.017
Cr	0.067	0.055	0.080	0.066
Cd	0.90	1.58	0.52	0.58

For most metals there was large difference in concentration of the same metals in different ginger sample (for example the concentration of Mg, Mn, Zn, Fe, and Cd of Illubabur and Tepi sample were higher than the other sample sites), this may be attributed to the difference in mineral concentration, the pH and organic matter content of the respective soil. The other probable reason is that the Tepi and Illubabur sample sites are near to the city and may be such differences are observed as a result of higher population and industrial activities in cities and municipalities which lead to higher production of assorted waste than in the rural settlements of Bombae and Hadaro.

4.5 Comparison of levels of metal in ginger of this study with literature values

The comparative study of the metal concentration of ginger determined in this study and reported values of other researchers are presented in Table 7. The Ca content of Ethiopian ginger is a bit higher than that of reported by Devi et al. (2008) and almost comparable to that of reported by

Ogunwandea and Olawore (2004). The level of Mg of this study is comparable to those studied in Nigeria (Ogunwandea and Olawore2004) but much lower than those in India (Devi et al. 2008).

Table 7. Comparison of determined metals concentration (mg/kg, dry mass basis) with reported values

Metal	Concentration (mg/kg)	Country	Reference
Ca	1700	India	(Devi et al. 2008)
	2610	Nigeria	(Ogunwandea and Olawore 2004)
	2001-2543	Ethiopia	This study
Mg	4210	Nigeria	(Ogunwandea and Olawore 2004)
	9200	India	(Devi et al. 2008)
	2700-4094	Ethiopia	This study
Mn	368	Nigeria	(Ogunwandea and Olawore 2004)
	313	India	(Devi et al. 2008)
	184-401	Ethiopia	This study
Fe	217	India	(Devi et al. 2008)
	19.4	Saudi Arabia	(Alwakeel 2008)
	41.8-89.0	Ethiopia	This study
Zn	72.53	India	(Devi et al. 2008)
	38.5-55.2	Ethiopia	This study
Ni	108	Nigeria	(Obiajunwa et al. 2002)
	5.61-8.40	Ethiopia	This study
Co	0.32	Saudi Arabia	(Al-Eed et al. 2002)
	2.04-7.58	Ethiopia	This study
Cu	4.47	India	(Devi et al. 2008)
	1.10-4.78	Ethiopia	This study
Cr	0.47	Pakistan	(Hashmi et al. 2005)
	0.5	Pakistan	(Hashmi et al. 2007)
	6.02 – 10.8	Ethiopia	This study
Cd	0.12	Nigeria	(Aiwonegbe and Ikhuoria 2007)
	0.07	Saudi Arabia	(Al-Eed et al. 2002)
	0.38-0.97	Ethiopia	This study
Pb	<0.021	Nigeria	(Aiwonegbe and Ikhuoria 2007)
	ND	Ethiopia	This study

ND: Concentration of the tested heavy metal was below the method detection limit.

Mn concentration in Ethiopian ginger is in the range of 184–401 mg/kg and its values reported by Devi et al. (2008) and Obiajunwa et al. (2002) lies in this range. The Fe content reported in Saudi Arabia (Alwakeel 2008) is less than the Fe content in Ethiopian ginger. However, the Fe content reported by Devi et al. (2008) is higher than the result of the present study. The mean concentration of Zn determined in this study is a bit lower than the value determined in India (Devi et al. 2008). The content of Zn in Ethiopian ginger is above the permissible limit set by FAO/WHO in edible plants (27.4 mg/kg). However, according to Bowen and Allaway, the range of Zn in agricultural products should be between 15 to 200 mg/kg (Jabeen et al. 2010).

The Ni concentration in Ethiopian ginger is lower than the Ni content determined in Nigeria (Obiajunwa et al. 2002). The Ni content in Ethiopian ginger is higher than the permissible limit set by FAO/WHO in edible plants (1.63 mg/kg). However, Ni toxicity in human is not a very common occurrence because its absorption by the body is very low (Jabeen et al. 2010).

From Table 7, one can see that the concentration of Cu in Ethiopian ginger is almost the same as the value reported in India (Devi et al. 2008). The Cr content of ginger reported by Hashmi et al. (2005) and Hashmi et al. (2007) are less than the present result. The concentration of Cr in Ethiopian ginger is higher than the permissible limit set by FAO/WHO in edible plants (0.02 mg/kg). Chronic exposure to Cr may result in liver, kidney and lung damage (Jabeen et al. 2010).

The level of Co obtained in Ethiopian ginger is 2.04–7.58 mg/kg, which is higher than the value reported in Saudi Arabia (Al-Eed et al. 2002). The Cd content in Ethiopian ginger is 0.38–0.97 mg/kg, which is higher than the Cd content determined in Nigeria (Aiwonegbe and Ikhuoria 2007) and Saudi Arabia (Al-Eed et al. 2002). The level of Cd in Ethiopian ginger is above the permissible limit set by FAO/WHO in edible plants (0.21 mg/kg). The results of this study reveal that the content of Pb in ginger is below the method detection limit and that assures the low Pb exposure of the farmlands.

The minimal risk levels for hazardous Pb and Cd metals through oral route and its acute effect are 0.0002 mg/kg per day for both metals (Al-Eed et al. 2002). Whereas the human need from ginger is very few grams per day, hence there is no risk from ginger used in the food.

4.6 Analysis of variance

T-tests and analysis of variance (ANOVA) are widely used statistical methods to compare group means. While the independent sample t-test is limited to comparing the means of two groups, the one-way ANOVA can compare the mean of more than two groups of sample. ANOVA uses the F statistic to compare whether the differences between sample means are significant or not (Bereton 2003).

In this study, ginger samples were collected from four different areas and the metal levels of each sample was analysed by FAAS. During the processes of sample preparation and analysis a number of random errors may be introduced in each aliquot and in each replicate measurement. The variation in sample mean of the analyte was tested by using ANOVA, whether the source for variation was from experimental procedure or heterogeneity among the samples (i.e. difference in mineral contents of soil, pH of soil, water, atmosphere; variation in application of agrochemicals like fertilizers, pesticides, herbicides etc. or other variations in cultivation procedures).

The ANOVA results (Table 8) showed that there exist statistically significant differences at 95% confidence level in mean concentrations of all the nine metals except Ni which is expected from variation in experimental procedure. The source for this significant difference between sample means may be the difference in mineral contents of soil or pH of soil which predict the extent of mineral absorption by ginger. For Ni the difference between samples mean is not significant. The variation among results of Ni in four ginger samples should not be attributed to anything more than random error in the analytical procedure.

4.7 Pearson correlation of metals between ginger and soil samples

In this study, to correlate the effect of one metal concentration on the concentration of the other metal in the same sample and to correlate the effect of the same metal concentration of soil to that of ginger, the Pearson correlation matrices using correlation coefficient (r) for the samples were used. The results are given in Tables 9 and 10. There is high positive correlation for Mg with (Ni, Zn, Mn, Fe and Cr), Cu with (Co and Cd), Zn with (Mn, Cr and Cd), Mn with Cd, Ni with Cr and moderate correlation for Cr with Cd; which may be arise from common anthropogenic or natural sources as well as from similarity in chemical properties. The high negative correlation between Ca and Cu indicate the large absorption of Ca may affect the absorption of Cu in ginger plant. The other metals have weak negative or positive correlation indicating that the presence or absence of one metal affect in lesser extent to the other. In soil samples, there is high positive correlation for all metals except Ca with (all metals except Mn), Co with (all metals except Zn). The low negative or positive correlation of Ca and Co with the other metals in the soil samples may be associated with chemical properties like insoluble carbonates.

Table 8. Analysis of variance (ANOVA) between and within ginger samples at 95% confidence level

Metal	Comparision	SD	df	$F_{calculated}$	$F_{critical}$	Remark
Ca	Between samples	255	3	4.96	4.07	Significant difference between sample means
	Within samples	51.4	8			
Mg	Between samples	650	3	14.2	4.07	Significant difference between sample means
	Within samples	45.6	8			
Cu	Between samples	1.58	3	8.32	4.07	Significant difference between sample means
	Within samples	0.19	8			
Zn	Between samples	9.96	3	5.89	4.07	Significant difference between sample means
	Within samples	1.69	8			
Mn	Between samples	100	3	13.8	4.07	Significant difference between sample means
	Within samples	7.27	8			
Ni	Between samples	1.38	3	3.14	4.07	No significant difference between sample means
	Within samples	0.44	8			
Fe	Between samples	21.7	3	5.09	4.07	Significant difference between sample means
	Within samples	4.28	8			
Co	Between samples	2.72	3	9.38	4.07	Significant difference between sample means
	Within samples	0.29	8			
Cr	Between samples	2.02	3	4.93	4.07	Significant difference between sample means
	Within samples	0.41	8			
Cd	Between samples	0.29	3	7.25	4.07	Significant difference between sample means
	Within samples	0.04	8			

where, SD - is standard deviation of between sample and within sample and df - is degree of freedom of between sample and within sample.

Table 9. Correlation matrices for metals in ginger sample (n = 4)

	Ca	Mg	Cu	Zn	Mn	Ni	Fe	Co	Cr	Cd
Ca	1									
Mg	0.315	1								
Cu	−0.907	−0.453	1							
Zn	−0.289	0.680	0.310	1						
Mn	0.032	0.683	0.120	0.924	1					
Ni	0.320	0.866	−0.623	0.311	0.232	1				
Fe	0.665	0.916	−0.711	0.443	0.592	0.783	1			
Co	−0.376	−0.426	0.731	0.286	0.370	−0.813	−0.438	1		
Cr	−0.312	0.751	0.002	0.655	0.399	0.781	0.439	−0.450	1	
Cd	−0.554	0.397	0.616	0.941	0.814	0.022	0.115	0.513	0.527	1

Table 10. Correlation matrices for metals in soil sample (n = 4)

	Ca	Mg	Cu	Zn	Mn	Ni	Fe	Co	Cr	Cd
Ca	1									
Mg	0.428	1								
Cu	0.445	0.996	1							
Zn	0.252	0.749	0.800	1						
Mn	0.706	0.937	0.948	0.737	1					
Ni	0.505	0.994	0.997	0.779	0.967	1				
Fe	0.461	0.987	0.997	0.839	0.953	0.995	1			
Co	0.403	0.313	0.393	0.806	0.477	0.395	0.464	1		
Cr	0.052	0.919	0.917	0.789	0.744	0.887	0.908	0.274	1	
Cd	0.282	0.942	0.910	0.512	0.808	0.905	0.875	−0.023	0.883	1

As shown in Table 11, one can see that the more levels of metals like Mg, Zn, Mn, Co, Cr and Cd in the soil, the more accumulation of corresponding metals in the ginger. This verifies that the dependence of metal concentration in the plant on metal concentration of respective soil.

Table 11. Pearson correlation coefficient for metals in ginger with soil sample (n = 4)

Metal	Ca	Mg	Cu	Zn	Mn	Ni	Fe	Co	Cr	Cd
r	−0.812	0.723	0.288	0.857	0.829	0.345	0.459	0.744	0.814	0.766

where, r is the Pearson correlation coefficient between metal level in ginger and metal level in soil.

5. CONCLUSION

An efficient digestion procedure for ginger and soil sample was developed and validated through standard addition (spiking) method and a good percentage recovery was obtained (100 ± 10%) for all minerals identified in ginger.

The levels of metals in ginger determined in this study could be put in the following order Mg (2700–4090 mg/kg) > Ca (2000–2540 mg/kg) > Mn (184–401 mg/kg) > Fe (41.8–89.0 mg/kg) > Zn (38.5–55.2 mg/kg) > Cr (6.02–10.8 mg/kg) > Ni (5.61–8.40 mg/kg) > Co (2.04–7.58 mg/kg) > Cd (0.38–0.97 mg/kg). The non-essential toxic heavy metal, Pb, was found to be below the method detection limit. The results of this work confess that ginger accumulates relatively higher amounts of Mg and Mn among the determined macro- and micronutrients, respectively.

The ANOVA results at 95% confidence level suggest that there were significant difference in the mean concentration of all metals except Ni between the four sampling areas which could be attributed to the difference in mineral contents of soil or pH of soil which predict the extent of mineral absorption by ginger. The soils of the study farmland were found to contain high levels of Fe followed by Mn, Ca, Mg, Zn, Cr, Co, Ni, Cu, Cd. The level of Pb in soils of all sampling sites was found to be below detection limit of the method used in this study.

In general, the levels of most of the metals in the studied soils were found to correlate positively with the levels found in the ginger.

ACKNOWLEDGEMENTS

The authors express their gratitude to the Department of Chemistry, Addis Ababa University, Addis Ababa, Ethiopia for providing laboratory facilities. YW is thankful to Department of Chemistry, Arbaminch College of Teachers Education, Arbaminch, Ethiopia for sponsoring his study.

COMPETING INTERESTS

The authors declare that they have no competing interests.

AUTHORS' CONTRIBUTIONS

YW contributed to the sampling and analysis. BSC contributed to the design of the study. YW and BSC contributed to the interpretation of data and drafting the manuscript. BSC contributed to critical revision of the paper. All authors have read the manuscript and gave their final approval for this version of the manuscript to be published.

REFERENCES

1. Aiwonegbe AE, Ikhuoria EU (2007) Levels of selected heavy metals in some Nigerian vegetables. Trends Appl Sci Res 2:76–79

2. Al-Eed MA, Assubaie FN, El-Garawany MM, El-Hamshary H, Eltayeb ZM (2002) Determination of heavy metal levels in common spices. J Appl Sci 17:87–98

3. Alwakeel SS (2008) Microbial and heavy metals accumulation of herbal medicines. Res J Microbiol 3:683–691

4. Aregahegn A, Chandravanshi BS, Atlabachew M (2013) Mineral contents of fruits of cactus pear (*Opuntia ficus indica*) grown in Ethiopia. Acta Hort (ISHS) 979:117–126

5. Asfaw N, Abegaz B (1995) Chemical constituents of the essential oils of *Zingiber officinale* Roscoe cultivated in Ethiopia. SINET Ethiop J Sci 18:133–135

6. Asfaw N, Demissew S (2009) Aromatic plants in Ethiopia. Shoma books, Addis Ababa, pp 27–185

7. Bereton RG (2003) Chemometrics: Data analysis for the laboratory and chemical plant. John Wiley and Sons, London, pp 197–206

8. Chen M, Ma LQ (2001) Comparison of three aqua-regia digestion methods for twenty florida soils. Soil Sci Soc Am J 65:491–499

9. Demirel S, Tuzen M, Saracoglu S, Soylak M (2008) Evaluation of various digestion procedures for trace element contents of some food materials. J Hazard Mater 152:1020–1026

10. Derbie A, Chandravanshi BS (2011) Concentration levels of selected metals in the leaves of different species of thyme (*T. schimperi* and *T. vulgaris*) grown in Ethiopia. Biol Trace Elem Res 141:317–328

11. Devi KN, Sarma HN, Kumar S (2008) Estimation of essential and trace elements in some medicinal plants by PIXE and PIGE techniques. Nucl Instr Methods Phys Res B 266:1605–1610

12. Dubale AA, Chandravanshi BS, Gebremariam KF (2015) Levels of major and trace metals in the leaves and infusions of *Croton Macrostachyus*. Bull Chem Soc Ethiop 29:11–26

13. Ebong GA, Akpan MM, Mkpenie VN (2008) Heavy metal contents of municipal and rural dumpsite soils and rate of accumulation by *Carica papaya* and *Talinum triangulare* in Uyo, Nigeria. E-J Chem 5:281–290

14. Endalamaw FD, Chandravanshi BS (2015) Levels of major and trace elements in fennel (*Foeniculum vulgari*Mill) fruits cultivated in Ethiopia. Springer Plus 4:5

15. Gaudino S, Galas C, Belli M, Barbizzi S, Zorzi P, Jacimovic R, Jeran Z, Pati A, Sansone U (2007) The role of different soil sample digestion methods on trace elements analysis: a comparison of ICP-MS and INAA measurement results. J Qual Comp Reli Chem Measur 12:84–93

16. Gebre A, Chandravanshi BS (2012) Levels of essential and non-essential metals in *Rhamnus prinoides* (Gesho) cultivated in Ethiopia. Bull Chem Soc Ethiop 26:329–342

17. Govindarajan VS (1982) Ginger – Chemistry, technology and quality evaluation: Part I. CRC Crit Rev 17:1–96

18. Grusak MA (2002) Enhancing mineral content in plant food products. J Am College Nutr 21:178–183

19. Hailemicheal G, Tilahun D, Etissa E (2008) Spices Research Achivements, Revised edn. Ethiopian Institute of Agricultural Research, Addis Ababa, pp 12–23

20. Harold HS, Leslie MK (2000) History of nutrition symposium: trace element nutrition and human health. J. Nutr Sci 130:4835–4845

21. Hashmi DR, Khan FA, Shaikh GH, Usmani TH (2005) Determination of trace metals in the vegetables procured from local markets of Karachi city by Atomic absorption spectrophotometery. J Chem Soc Pak 27:353–357

22. Hashmi DR, Ismail S, Shaikh GH (2007) Assessment of the level of trace metals in commonly edible vegetables locally available in the markets of Karachi city. Pak J Bot 39:747–751

23. Huang L, Bell RW, Dell B, Woodward J (2004) Rapid nitric acid digestion of plant material with an open-vessel microwave system. Commun Soil Sci Plant Anal 35:427–440

24. Jabeen S, Shah MT, Khan S, Hayat MQ (2010) Determination of major and trace elements in ten important folk therapeutic plants of Haripur basin, Pakistan. J Med Plants Res 4:559–566

25. Jolad SD, Lantz RC, Solyom AM, Chen GJ, Bates RB, Timmermann BN (2004) Fresh organically grown ginger (*Zingiber officinale*): composition and effects on LPS-induced PGE_2 production. Phytochemistry 65:1937–1954

26. Jung MC (2008) Heavy metal concentrations in soils and factors affecting metal uptake by plants in the vicinity of a Korean Cu-W mine. Sensors 8:2413–2423

27. Kitata RB, Chandravanshi BS (2012) Concentration levels of major and trace metals in onion (*Allium cepa* L.) and irrigation water around Meki Town and Lake Ziway, Ethiopia. Bull Chem Soc Ethiop 26:27–42

28. Lokeshwari H, Chandrappa GT (2006) Impact of heavy metal contamination of bellandur lake on soil and cultivated vegetation. Curr Sci 91:622–627

29. Mekebo D, Chandravanshi BS (2014) Levels of essential and non-essential metals in linseed (*Linum usitatissimum*) cultivated in Ethiopia. Bull Chem Soc Ethiop 28:349–362

30. Minerals-Learn (2010) The role of minerals in maintaining a healthy body. Available at: http://www.expert-nutrition.com/minerals.html, accessed on May 22, 2010.

31. Mishra P (2009) Isolation, spectroscopic characterization and molecular modeling studies of mixture of curcuma longa, ginger and seeds of fenugreek. Int J PharmTech Res 1:79–95

32. Mitra S (2003) Sample Preparation Techniques in Analytical Chemistry. John Wiley and sons, Inc., Hoboken, pp 227–244

33. NebGuide (2015) Upper safe levels of intake for adults: vitamins and minerals. Available at: http://www.ianrpubs.unl.edu/pages/publicationD.jsp?publicationId=295, accessed on February 13, 2015

34. Obiajunwa EI, Adebajo AC, Omobuwajo OR (2002) Essential and trace element contents of some Nigerian medicinal plants. J Radioanal Nucl Chem 252:473–476

35. Ogunwandea IA, Olawore NO (2004) Heavy trace metals and macronutrients status in herbal plants of Nigeria. Food Chem 85:67–71

36. Pandey J, Pandey U (2009) Accumulation of heavy metals in dietary vegetables and cultivated soil horizon in organic farming system in relation to atmospheric deposition in a seasonally dry tropical region of India. Environ Monit Assess 148:61–74

37. Parthasarathy VA, Chempakam B, Zachariah TJ (2008) Chemistry of Spices. CAB International, Wallingford, UK, pp 70–93

38. Uwah EI, Ndahi NP, Ogugbuaja VO (2009) Study of the levels of some agricultural pollutants in soils, and water leaf (*Talinum triangulare*) obtained in Maiduguri, Nigeria. J Appl Sci Environ Sanit 4:71–78

39. Weldegebriel Y, Chandravanshi BS, Wondimu T (2012) Concentration levels of metals in vegetables grown in soils irrigated with river water in Addis Ababa, Ethiopia. Ecotoxicol Environ Saf 77:57–63

40. Wilson B, Braithwaite A, Pyatt FB (2005) An evaluation of procedures for the digestion of soils and vegetation from areas with metalliferous pollution. Toxicol Environ Chem 87:335–344

41. Zingiber officinale (2010) Available at: http://www.allianceingredients.com/pdfdocs/ZINGIBER_OFFICINALE.PDF, accessed on February 25, 2010

42. Zubillaga MS, Bressan E, Lavad RS (2008) Heavy metal mobility in polluted soils: effect of different treatments. Am J Environ Sci 4:620–624

Chapter 5

Levels of essential and non-essential metals in edible mushrooms cultivated in Haramaya, Ethiopia

Medhanye Gebrelibanos[1*], Negussie Megersa[2] and Abi M. Taddesse[3]

[1]Department of Industrial Chemistry, School of Biological and Chemical Sciences and Technology, Addis Ababa Science and Technology University, P. O. Box 16417, Addis Ababa, Ethiopia.

[2]Department of Chemistry, College of Natural Sciences, Addis Ababa University, P. O. Box 1176, Addis Ababa, Ethiopia.

[3]Department of Chemistry, College of Natural and Computational Sciences, Haramaya University, P. O. Box 138, Haramaya, Ethiopia.

ABSTRACT

Background

Fruiting bodies of two oyster mushroom (*Pleurotus ostreatus* and *Pleurotus florida*) species and their respective substrate samples cultivated and collected from the Mushroom Research, Production and Training Laboratory of the Haramaya University, Ethiopia, were analyzed for essential and non-essential metals, *viz.*, Cu, Fe, Zn, Mn, Cd and Pb contents by flame atomic absorption spectrometry after digestion employing dry, wet and microwave digestion techniques. Concentrations of the analyzed metals were expressed on a dry weight basis (mg/kg).

Results

The mean metal concentration (mg/kg in dry weight) in the mushroom and respective substrate samples, respectively; for *Pleurotus ostreatus* were: Cu (51.19, 48.97), Fe (220.87, 299.51), Zn (89.68, 62.55) and Mn (47.55, 58.32). The mean metal concentration (mg/kg in dry weight) in mushroom and the respective substrate samples, respectively, for *Pleurotus florida* were: Cu (53.56, 51.87), Fe (243.92, 265.23), Zn (95.26, 64.25) and Mn (41.29, 62.55). However, Cd and Pb was not detected in both mushroom and substrate samples. The bio-accumulation factor of the heavy metals was calculated and a highest accumulation of Cu and Zn from substrate was observed for all the analyzed mushrooms samples. The observed results were also compared with the reported results.

Conclusions

In this study, the highest concentrations of Cu, Fe, Zn and Mn were confirmed. These elements generally are essential for humans, and may not possibly cause health problems even if mushrooms are consumed regularly. Furthermore, the quantities of the metals found in the samples were within the limits recommended by international organizations like the WHO/FAO Expert Committee on Food Additives. On the other hand, the amounts of Cd and Pb were found to be very low. These elements are toxic to humans, especially at higher doses; however, the concentrations of these non-essential metals were below detectable limit.

KEYWORDS

Edible mushrooms, Substrate Fruiting body, Sample digestion, Essential metals, Non-essential metals BCF

1. BACKGROUND

Mushrooms are fungi fruit-bodies that spontaneously grow in forests and farmlands in large quantities after the rainy seasons. The natural substrates of the mushrooms include logs of woods, decomposed agro- and animal wastes, and soil where nutrients are available through external digestion and absorption by the mycelium. There are edible and poisonous mushrooms and both categories possess nutritional and medicinal values (Okwulehie and Ogoke 2013). The consumption of wild edible mushrooms is increasing from time to time, in the developed world, mainly due to their contents of proteins as well as trace minerals (Agrahar-Murugkar and Subbuakshmi 2005). Besides, wild growing and cultivated mushrooms are considered popular food in several countries, mainly in Europe and Asia, because of their favorite delicacy (Kalac 2010; Zhu et al. 2011). Furthermore, mushrooms have been reported as therapeutic foods; useful in preventing from diseases and exhibiting varied biological properties such as antibacterial, antimutagenic, antitumoral and antiviral activities (Garcia-Lafuente et al. 2011; Schillaci et al.2013). These functional characteristics are mainly because of their chemical composition (Manzi et al.2001). In this regard, fruiting bodies of the mushrooms, on a dry weight basis, contain about 39.9 % carbohydrate, 17.5 % protein and 2.9 % fats, the rest being minerals (Demirbas 2001).

Living organisms require varying amounts of "heavy metals" including iron, copper, manganese and zinc, that are essential and required by humans as they play important roles in biological systems, though excessive levels may result in damaging effects to the organism. Other heavy metals such as cadmium and lead are toxic and non-essential and their accumulation over time in human bodies could cause serious illnesses (Obodai et al. 2014). Accurate and adequate food composition data are invaluable for estimating the adequacy of intakes of essential metals and assessing exposure risks from intake of toxic non-essential metals (Soylak et al. 2005). For example, the bioavailability of iron in mushrooms is considerably high and human body can absorb up to 90 % of the available form (Kalac and Svoboda2000). The recommended dietary allowances (RDA) are 0.9, 8.0–18.0, 8.0–11.0 and 1.8–2.3 mg per day for Cu, Fe, Zn and Mn, respectively. This level is sufficient to meet the nutrient requirement of nearly all (97 to 98 %) healthy individuals, in a particular life stage and gender group (Ross et al. 2011).

Metal concentrations in mushrooms are considerably higher than those in agricultural crops, vegetables and fruits. This suggests that mushrooms possess effective mechanisms that enable them to readily take up some metals from the ecosystem compared to the green plants growing in similar localities (Svoboda et al. 2000). Reported study has indicated that concentrations of metals in macrofungi depend on the physiology of the species and particularly on the ecosystem pattern (Sesli et al. 2008). In addition, accumulation of metals in the mushrooms was found to be species metabolism-dependent and also strongly affected by the chemical composition of the substrate from which the mushrooms get their nutrients (Radulescu et al. 2010). As the macrofungi are integral part of the forest ecosystems, sometimes the substrate-to-mycelium transfer of metals depend on the relationship between mycelium and symbiotic plant species affecting element absorption and translocation (Yoshida and Muramatsu 1997).

Furthermore, accumulation of heavy metals in macrofungi has been noted to be affected by environmental and fungal factors. Environmental factors, such as organic matter contents, pH, metal concentrations in soil or substrate and fungal factors, such as species of mushroom, morphological part of fruiting body, development stages, age of mycelium and biochemical composition are known to affect metal accumulation in macrofungi (Kalac and Svoboda 2000; Isiloglu et al. 2001).

The importance of mushrooms in the ecosystem can be described by their ability to biodegrade the substrate and therefore use the wastes of agricultural production (Sarikurkcu et al. 2011). In several countries, there is a well-established consumer acceptance of cultivated mushrooms, such as *Agaricus bisporus*, *Pleurotus* species, *Lentinus edodes* and others (Diez and Alvarez 2001). Kues and Liu (2000) reported that the *Pleurotus* species are well known edible mushrooms in different parts of the world while other workers stated that oyster mushroom (*Pleurotus florida*) cultivation is gaining popularity due to low cost technology and easy availability of the various substrates for its cultivation (Shukla and Biswas (2000)). *Pleurotus* species, as primary wood rot fungi, have the capacity to colonize different types of agricultural wastes as substrates (Zadrazil and Dube 1992). Additionally, mushrooms are commonly cultivated on single substrates, commonly wheat or paddy straw (Ragunathan and Swaminathan 2003).

The mushroom cultivation practices and consumption in Ethiopia is still very low compared to other countries in Europe and Asia. However, mushroom consumption is increasing due to the intense advertising of its numerous benefits. Besides their nutritional values, mushrooms possess medicinal proprieties in prevention of some diseases, such as hypertension and cancer (Manzi et al. 2001). Although there were several studies carried out in several countries on metals concentrations in the fruiting bodies of edible wild grown and cultivated mushrooms (Tuzen et al. 2007; Sesli et al. 2008; Baslar et al. 2009; Radulescu et al. 2010; Sarikurkcu et al. 2011; Zhu et al. 2011; Sen et al. 2012; Okwulehie and Ogoke 2013; Obodai et al. 2014; Dulay et al. 2015), however, such studies are still very scarce in Ethiopia. Therefore, initiation of such study for better understanding of the types of mushrooms cultivated and consumed in the country in relation to the essential and non- essential metals is very much demanding and timely.

Since heavy metals could enter the food chain, for example, as a result of uptake by edible mushrooms, it is necessary to determine the levels of essential and non-essential metals and report possible contamination level that could be the cause for health hazard. However, there have been no such reports, on the essential and non-essential metal levels in oyster mushroom samples cultivated in Ethiopia. In the light of the above, the elemental contents of six essential and non-essential metals (Cu, Zn, Fe, Mn, Cd, and Pb) in two different edible oyster mushroom species cultivated in Haramaya, Ethiopia, on wheat straw substrate were investigated.

2. METHODS

2.1 Description of the sampling area

Cultivation of oyster mushrooms (*Pleurotus* spp.) was carried out at the Mushroom Research, Production and Training Laboratory of the Haramaya University, Ethiopia; located at 42°3’E longitude, 9°26’N latitude and altitude of 1980 m above the sea level (AUA 1996). Temperature of the experimental room was found to vary from 12.96 to 25.72 °C, while the relative humidity ranging from 57.67 to 91.93 %.

2.2 Oyster (*Pleurotus* spp.) mushroom cultivation

2.2.1 Source of mushroom strain

Pure cultures of two edible oyster mushroom species; namely; *Pleurotus ostreatus* and *Pleurotus florida,*were obtained from the cultivation center.

The cultures were initially multiplied on sterilized potato dextrose agar (PDA) medium to get enough stock culture of the pure species needed to prepare the required quantity of spawn (Sara 2007). Stock culture was grown by transferring the PDA cultures into the flasks containing maize seeds as described below (subsection Spawn preparation).

2.3 Spawn preparation

2.3.1 Preparation and sterilization of spawn substrate

Mother spawn was prepared according to the procedure described by Singh and Chaube (1995). Accordingly, 9 kg maize seeds were boiled in 15 L water for 15 min and allowed to remain soaked in the hot water for another 25 min. The water was then drained off and the maize seeds were kept in a sieve overnight to dry. The next day, 120 g calcium sulfate and 30 g calcium carbonate were mixed with each boiled maize seeds, in order to maintain the pH close to neutrality and reduced adhesion of the maize seeds (Smith and Margarel 1995). Thereafter, the supplemented maize seeds were filled in half or one liter sterilizable bottles (225 or 450 g/bottle), plugged with non-absorbent cotton and sterilized in autoclave at 121 °C for half an hour. Subsequently, the bottles were cooled and then inoculated with pure stock culture by taking a piece of agar with mycelium. At this stage, the culture and the prepared maize seeds were thoroughly mixed in order to distribute the mycelium uniformly which were incubated at 25 °C. Finally, the maize seeds were fully covered with mycelium, and then the bottles were used as a mother spawn.

2.3.2 Inoculation of the spawn substrate

Inoculation of spawn substrate was performed by transferring small amount of agar with mycelium, from stock culture, to the flask containing the substrate under aseptic condition over the flame. This was followed by thoroughly mixing the mycelium with the maize seeds in the flasks and the flask was then incubated at 25 °C, in an incubator. After seven days of incubation, the content of the flasks was shaken to facilitate uniform mixing of the mycelium over the maize seeds. The flask was then incubated until all the maize seeds fully covered with mycelium in about 15 days (Sara 2007).

2.4 Substrate preparation and cultivation

The substrate, wheat straw, was collected from the farmlands of the mushroom research center of the Haramaya University. It was then undergone the following experimental steps: initially the wheat straws were chopped to approximately 8 – 12 cm size pieces and filled in the gunny bags, and then soaked in 2 % aqueous formalin solution for about 18 h. The bags were taken out of the solution and allowed to drain the excess solution for 4 – 6 h. The soaked pieces of the wheat straw materials were then checked for remaining excess solution by pressing between the palms. When there is no dripping of solution, the chopped wheat straw material was considered 'ready for spawning' (Sara 2007). The wet substrate materials were then spread over a clean alcohol swabbed polyethylene sheet and inoculated with the spawn; prepared at the rate of 3 % on wet basis. The resulting spawn was thoroughly mixed with the substrate and then filled into the plastic bags (about 2 kg per bag). Following spawning, the bags were kept about 15 cm apart in a crop room; with the temperature maintained around 25 – 30 °C and humidity ranging from 70 to 90 %. The temperature and humidity of the cropping room were controlled to the required ranges using a heater and spraying water to the walls and floors of the room. During cropping or harvesting period, the bags were sprinkled with water twice a day (Sara 2007). Finally, after cultivation period of 25 – 35 days, the fully matured mushroom species, grown on the substrate (wheat straw), were harvested and collected for further treatment. This experiment was performed in triplicate (Fig. 1).

2.5 Sample collection and preparation

For each fruiting body of the mushrooms and their respective substrates, a total of three samples were collected from the cultivation center. The samples were collected during the months of December 2014 and January 2015.

Fruiting bodies of the mushroom samples collected from the center were thoroughly cleaned, cut and oven dried at 60 °C for 4 – 6 h. The resulting dried samples were then homogenized using agate homogenizer and stored in pre-cleaned polyethylene bottles for subsequent analysis. The substrate samples, on the other hand, were collected from the surface (0 – 15 cm in depth), where the mushroom samples were grown. They were oven dried at 70 °C until the entire moisture contained within the sample was removed. After drying, the substrate samples have been homogenized to a fine powder and stored in a pre-cleaned polyethylene bottles for subsequent analysis.

Figure 1. Spawn and substrate preparation, and mushroom cultivation

2.6 Digestion procedures

2.6.1 Dry ashing

Dried mushroom and substrate samples (0.5 g) were first placed onto a porcelain crucible. The furnace temperature was slowly increased from room temperature to 450 and 500 °C for the mushroom and substrate samples, respectively, in 1 h. Both samples were ashed for about 4 h until a white or grey ash residue was obtained. The residues were then dissolved in 1 mL HNO_3 and 2 mL HNO_3 for mushroom and substrate samples, respectively, and the resulting solutions were transferred to a 50 mL volumetric flask and made up to a volume with deionized water. Preparation of the blank control, in the absence of analytes, was also carried out in a similar manner.

2.6.2 Wet digestion

Wet digestion of the mushroom and substrate samples, 0.5 g each, was performed separately in a 100 mL beaker, inside the fume hood. The optimum oxi-acid mixture ratio for mushroom digestion was 2:1:1 (*v/v/v*) (HNO_3:H_2SO_4:H_2O_2), in a total volume of 12 mL, for 3 h at 150 °C. On the other hand, for the substrate samples the optimum digestion conditions were found to be 3 mL HNO_3 and 9 mL HCl (*aqua regia*) and a digestion time of 1 h at 100 °C. Then, the clear solutions for both samples were made up to the final volume of 50 mL with deionized water. Digestion of the blank control, in the absence of analytes, was also carried out in a similar manner.

2.6.3 Microwave digestion

Accurately weighed, 0.5 g of both the mushroom and substrate samples were separately digested, each for 2 min. Digestion of the mushroom samples was performed in the acid mixture composed of 6 mL of HNO_3, 3 mL H_2SO_4 and 3 mL of H_2O_2 in a power of 250 W. Similarly, for the substrate samples, aqueous acid mixture composed of 6 mL HCl and 3 mL HNO_3, in 200 W power of microwave digestion system. Finally, the resulting clear solutions were diluted to the final volume of 50 mL with deionized water. Digestion of the blank control, in the absence of analytes, was also carried out in a similar manner.

2.7 Evaluation of the method performance and validation of the analytical method

Standard solutions for calibration were prepared, at five points, from 10 mg/L intermediate solutions of the respective metals. The instrument was calibrated using a series of working standards. Based on the results obtained from this experimental procedure, concentrations of the various metals in the digests and the corresponding correlation coefficients were determined. The precision of the results were also evaluated based on the standard deviation of the results of triplicate samples ($n = 3$), analyzed under the same conditions.

Limit of detection of the analytical method, LOD, is the lowest analyte concentration that can be detected from the statistical fluctuations of the blank, usually corresponds to the signal that is three times the standard deviation of the blank ($LOD = 3*S_b$, where S_b is the standard deviation (SD) of the blanks) (Miller and Miller 2010). LOD for each element was determined from the standard deviations and calculated based on twelve blank measurements. To this end, twelve blank samples were digested following the same procedures used for the samples and each of the blank samples was analyzed for the metal concentrations, i.e., Cu, Fe, Zn, Mn, Cd and Pb.

The limit of quantification, LOQ, is described as the lowest concentration level at which the measured values could be determined quantitatively and most often defined as 10 times the signal/noise ratio. In this study, LOQ was obtained from triplicate analyses of the blanks, which were digested following the same procedures, both for the mushroom and substrate samples. The LOQ was then calculated using the relation; $LOQ = 10*S_b$, $n = 3$, (Miller and Miller 2010).

Similarly, the percent recovery (%R) was determined by spiking the samples with a standard solution of 100 mg/L of the analyte metals. From these solutions based upon the amount that makes the concentration of the final solutions to 0.25 mg/L (Cu), 1.50 mg/L (Fe), 0.10 mg/L (Zn) and 1.00 mg/L (Mn) were added to 0.5 g of both mushroom and the respective substrate samples. Then, the contents were digested following the developed digestion procedures for all the samples. The digested spiked samples were finally analyzed for their respective metal contents. For both mushroom and substrate samples, triplicate digestions were performed and from the recovered quantity of each analyte, percent recoveries were calculated using the equation given below:

$$\%\ \text{Recovery} = \frac{\text{Amount After Spike} - \text{Amount Before Spike}}{\text{Amount Added}} \times 100$$

2.8 Statistical analysis

All experimental results were presented as means ± SD of three parallel measurements. The one way ANOVA and Fisher's least significant difference (LSD) was calculated using Microsoft Excel. Significant difference between treatment means was evaluated using analysis of variance (ANOVA) and for comparison of the means of the treatments, the LSD test was used at $p < 0.05$ significance level (Miller and Miller 2010).

3. RESULTS AND DISCUSSION

3.1 Evaluation of the analytical method

Calibration graphs for the six metallic elements, considered in this study, were constructed from the standard solutions at five points, ranging in concentration from 0.5 to 10 mg/L. All the elements exhibited linear relationships of the instrumental response and the solutions containing the metals with insignificant intercepts and correlation coefficients of 0.997 or better.

Observed detection and quantification limits for the metals under the study (in mg/kg) are given in Table 1. It has also been noted that the method detection limit for each element is above the instrument detection limit, for all the metals under analysis, determinations of which were carried out on dry weight basis.

Table 1. LOD and LOQ for the method of analysis of the mushroom and substrate samples

Element	IDL[a] (mg/L)	LOD for mushroom (mg/L)	LOD for substrate (mg/L)	LOQ for mushroom (mg/L)	LOQ for substrate (mg/L)
Cu	0.020	0.053	0.074	0.175	0.248
Fe	0.030	0.345	0.176	1.149	0.588
Zn	0.005	0.017	0.016	0.055	0.053
Mn	0.001	0.070	0.195	0.233	0.649
Cd	0.005	0.013	0.014	0.042	0.046
Pb	0.100	0.385	0.231	1.283	0.770

[a]Instrumental detection limit

Precision is the most often quoted term to express the extent of errors in the given analytical results. In this regard, the major concern is critical evaluation of the agreement between the measured values of the set of results for the same quantity (Miller and Miller, 2010). It is usually expressed by the relative standard deviation. In this study, the relative standard deviation determined for each target analyte was found to be less than 10 %, Table 2, which indicates the acceptable variability of the analytical results (Eurachem/CITAC 2003; EC 2010).

3.2 Percent recovery of the analytical method

Percent recovery analysis is used for method validation which is the process used for evaluating whether the analytical method is acceptable or not for its intended purpose. Due to the absence of the certified reference materials both for mushroom and substrate samples in our laboratory, validity of the optimized digestion procedures was assured by spiking the samples with the standard solution of known concentration of the analyte metals. As has also been shown in Table 3, the percentage recovery for the mushroom samples lied in the range 91.96 ± 3.32 % – 103.84 ± 4.82 %, which are within the acceptable range for all metals (EC 2010). Similarly, for the substrate samples, Table 4, the recoveries obtained were found to vary from 87.76 ± 1.25 to 104.52 ± 0.72 %, which are again within the acceptable range for all metals (EC 2010).

3.3 Metal contents in the mushrooms

The mean concentrations of essential and non-essential metals are presented in Table 2 for the samples digested by the three different digestion methods. All metal concentrations were determined on a dry weight basis as mg/kg. Among the investigated six essential and non-essential metal elements, all were found to be above the method detection limit, except for cadmium and lead, which were not detected in the mushroom (*Pleurotus ostreatus* and *Pleurotus florida*) and their respective substrate samples.

Table 2. Mean concentration ($n = 3$, x¯¯¯x¯ ± SD mg/kg dry weight) and %RSD of metals in mushroom and substrate samples

Digestion methods	Species	Cu		Fe		Zn		Mn		Cd	Pb
		Mean ± SD	%RSD	Mean ± SD	%RSD	Mean ± SD	%RSD	Mean ± SD	%RSD		
Dry ashing	PO	49.65 ± 1.75	3.53	200.12 ± 6.64	3.32	87.07 ± 0.75	0.87	46.08 ± 2.33	5.05	ND	ND
	Substrate	45.38 ± 2.86	6.30	287.18 ± 5.88	2.05	60.94 ± 0.72	0.18	56.28 ± 3.75	6.66	ND	ND
	PF	52.68 ± 3.51	6.66	234.60 ± 6.64	2.83	93.60 ± 1.83	1.95	39.36 ± 2.33	5.92	ND	ND
	Substrate	49.51 ± 4.29	8.66	257.70 ± 6.66	2.59	63.24 ± 1.11	1.75	60.61 ± 3.75	6.19	ND	ND
Wet digestion	PO	51.67 ± 3.04	5.88	226.05 ± 6.64	2.94	90.68 ± 0.96	1.06	47.04 ± 4.66	9.90	ND	ND
	Substrate	50.33 ± 5.15	10.24	301.28 ± 4.44	1.47	62.67 ± 0.20	0.32	58.44 ± 6.49	11.11	ND	ND
	PF	53.70 ± 1.75	3.27	245.21 ± 6.64	2.71	94.43 ± 1.83	1.93	41.67 ± 4.66	11.17	ND	ND
	Substrate	52.81 ± 3.78	7.16	266.67 ± 5.88	2.20	64.05 ± 0.40	0.62	62.77 ± 3.75	5.97	ND	ND
Microwave digestion	PO	52.25 ± 1.81	3.46	236.43 ± 6.71	2.84	91.29 ± 0.93	1.02	49.53 ± 2.32	4.68	ND	ND
	Substrate	51.20 ± 4.79	9.35	310.08 ± 6.71	2.17	64.04 ± 0.20	0.32	60.24 ± 4.02	6.67	ND	ND ND
	PF	54.34 ± 1.81	3.33	251.94 ± 6.71	2.66	97.76 ± 0.54	0.55	42.84 ± 2.32	5.41	ND	ND
	Substrate	53.29 ± 3.14	5.88	271.32 ± 6.71	2.44	65.45 ± 0.41	0.62	64.26 ± 4.02	6.25	ND	ND
Mean total	PO	51.19 ± 1.36	2.67	220.87 ± 18.70	8.47	89.68 ± 2.28	2.54	47.55 ± 1.78	3.74	ND	ND
	Substrate	48.97 ± 3.14	6.41	299.51 ± 11.55	3.86	62.55 ± 1.55	2.48	58.32 ± 1.98	3.40	ND	ND
	PF	53.56 ± 0.82	1.54	243.92 ± 8.74	3.58	95.26 ± 2.20	2.31	41.29 ± 1.77	4.29	ND	ND
	Substrate	51.87 ± 2.06	3.97	265.23 ± 6.92	2.61	64.25 ± 1.12	1.74	62.55 ± 1.84	2.93	ND	ND

ND Not Detected, *PO* Pleurotus ostreatus, *PF* Pleurotus florida

Table 3. Values of the recovery analysis (%R ± SD, $n = 3$) for mushroom samples

Metals		Dry ashing		Wet digestion		Microwave digestion	
		PO	PF	PO	PF	PO	PF
Cu	C^M	49.65	52.68	51.67	53.70	52.25	54.34
	AR	74.46 ± 2.41	78.64 ± 1.21	75.16 ± 3.62	77.94 ± 5.25	75.24 ± 8.30	78.37 ± 6.27
	%R	99.24 ± 0.96	103.84 ± 4.82	93.96 ± 1.45	96.96 ± 2.10	91.96 ± 3.32	96.12 ± 0.63
Fe	C^M	200.12	234.60	226.05	245.21	236.43	251.94
	AR	340.07 ± 3.86	382.16 ± 1.46	380.47 ± 5.26	392.26 ± 5.26	375.97 ± 13.43	395.35 ± 11.63
	%R	93.30 ± 2.57	98.37 ± 0.97	102.95 ± 3.50	98.03 ± 3.50	93.03 ± 0.89	95.61 ± 0.78
Zn	C^M	87.07	93.60	90.68	94.43	91.29	97.76
	AR	97.05 ± 1.07	103.28 ± 0.51	100.17 ± 0.88	104.29 ± 0.51	100.58 ± 0.89	107.04 ± 1.81
	%R	99.80 ± 1.07	96.80 ± 0.51	94.90 ± 0.88	98.60 ± 0.51	92.90 ± 0.88	92.80 ± 1.81
Mn	C^M	46.08	39.36	47.04	41.67	49.53	42.84
	AR	148.15 ± 3.26	136.04 ± 1.23	148.86 ± 8.09	141.74 ± 4.45	141.90 ± 4.64	137.23 ± 9.64
	%R	102.07 ± 3.26	96.68 ± 1.23	101.82 ± 0.81	100.07 ± 4.45	92.37 ± 0.93	94.39 ± 0.96

C^M Concentration in the Mushroom, *AR* Amount Recovered, *%R* Percent Recovery

Table 4. Values of the recovery analysis (%R ± SD, $n = 3$) for substrate samples

Metals		Dry ashing		Wet digestion		Microwave digestion	
		PO	PF	PO	PF	PO	PF
Cu	C^S	45.38	49.51	50.33	52.81	51.20	53.29
	AR	69.59 ± 2.41	72.37 ± 3.20	74.46 ± 4.35	77.24 ± 3.62	77.33 ± 1.81	75.23 ± 3.14
	%R	96.84 ± 0.96	91.44 ± 1.28	96.52 ± 1.74	97.72 ± 1.45	104.52 ± 0.72	87.76 ± 1.25
Fe	C^S	287.18	257.70	301.28	266.67	310.08	275.19
	AR	430.14 ± 5.83	405.72 ± 6.36	451.18 ± 3.86	414.98 ± 1.46	457.36 ± 6.71	418.61 ± 11.63
	%R	95.31 ± 3.89	98.68 ± 4.24	99.93 ± 2.57	98.87 ± 0.97	98.19 ± 4.48	95.61 ± 0.78
Zn	C^S	60.94	63.24	62.67	64.05	64.04	65.45
	AR	70.90 ± 1.02	72.90 ± 2.27	71.90 ± 0.70	74.01 ± 1.68	73.90 ± 1.08	74.96 ± 0.73
	%R	99.60 ± 1.02	96.60 ± 2.27	92.30 ± 0.69	99.60 ± 1.68	98.60 ± 1.08	95.10 ± 0.73
Mn	C^S	56.28	60.61	58.44	62.77	60.24	64.26
	AR	153.13 ± 6.53	161.68 ± 6.53	155.27 ± 4.45	160.97 ± 4.45	156.63 ± 6.96	163.32 ± 9.28
	%R	96.85 ± 0.65	101.07 ± 0.65	96.83 ± 4.44	98.2 ± 4.44	96.39 ± 0.69	99.06 ± 0.92

C^S Concentration in the Substrate, *AR* Amount Recovered, *%R* Percent Recovery

Copper is essential metal, which serve as a constituent of some metalloenzymes, and is required in hemoglobin synthesis and catalysis of metabolic growth (Silvestre et al. 2000). It was detected both in the fruiting body of the mushroom samples with varying concentrations. *Pleurotus florida* contained the highest amount (53.56 ± 0.82 mg/kg) of copper while least concentration was exhibited in the *Pleurotus ostreatus* sample (51.19 ± 1.36 mg/kg) (Table 2). The average copper content of the analyzed mushroom samples was 52.38 mg/kg, which are above the safe limit of 40 mg/kg set by WHO (1982) in foods. According to the review reported by Kalac (2010), copper concentrations accumulated in mushroom species were usually varied from 20.0 to 100.0 mg/kg. Other workers have also presented copper contents of the mushroom samples in the literature to be in the ranges: 16.1 – 144.9 mg/kg (Elekes et al. 2010), 2.6 – 72.4 mg/kg (Sen et al. 2012), 1.8 – 22.3 mg/kg (Okwulehie and Ogoke 2013), 5.0 – 83.0 mg/kg (Uzun et al. 2011) and 13.7 – 182.4 mg/kg (George et al. 2014). The results obtained, in the current study, indicated that copper content of the investigated mushroom samples were found to be comparable with those reported in the literature.

The iron content of the investigated mushroom samples of this study ranged from 220.87 ± 18.70 mg/kg in*Pleurotus ostreatus* to 243.92 ± 6.92 mg/kg in *Pleurotus florida* (Table 2). The average concentration of iron in the analyzed mushroom samples was 232.40 mg/kg, which was above the safe limit of 15 mg/kg set by WHO (1982). Based on the results from papers published until 1999, reported iron content was found to vary between 30.0 and 150.0 mg/kg (Kalac and Svoboda 2000). Similarly, iron contents determined in mushroom samples, during the recent years, were reported in the range of 44.4 – 125.1 mg/kg (George et al. 2014), 101.6 – 543.8 mg/kg (Radulescu et al. 2010), 1322.4 – 1596.0 mg/kg (Naresh et al. 2012), 139.0 – 1714.0 mg/kg (Sarikurkcu et al. 2011), 43.7 – 65.1 mg/kg (Quarcoo and Adotey 2013) and 8.25 – 58.25 mg/kg (Udochukwu et al. 2014), which could possibly be due to increased use of iron associated with increased industrialization. Thus, amount of iron obtained, in the present study, are in good agreement with the reported values in the literature. It is known that adequate iron in a diet is very important in order to decrease the incidence of anemia.

Zinc is one of the most important mineral needed by our body systems due to the fact that it is highly associated with protein and carbohydrate rich foods. Zinc is also used in medicines that treat rashes, acne, dandruff and athlete's foot (Okwulehie and Ogoke 2013). It has biological significances for living organisms and mushrooms are known as good zinc accumulators

(Isiloglu et al. 2001). Zinc was found in all the analyzed fruiting body of the mushrooms with concentration ranging from 89.68 ± 2.28 mg/kg in*Pleurotus ostreatus* and 95.26 ± 2.20 mg/kg in *Pleurotus florida* (Table 2). The average zinc content of the analyzed mushroom samples was 92.47 mg/kg. The values for zinc in the investigated mushroom samples were above the permissible limit of 60 mg/kg recommended values in foods (WHO 1982). Zinc concentrations of edible mushroom samples in the literature have been reported to be: 26.7 – 186.0 mg/kg (Genccelep et al. 2009), 19.0 – 49.7 mg/kg (George et al. 2014), 5.7 – 97.5 mg/kg (Sen et al.2012), 29.0 – 146.0 mg/kg (Sarikurkcu et al. 2011), 27.8 – 70.9 mg/kg (Okwulehie and Ogoke 2013) and 6.4–27.33 mg/kg (Udochukwu et al. 2014). This reveals that the findings of this study are in good agreement with the results reported for similar studies.

Manganese is an essential metal needed for biological systems such as metalloproteins (Unak et al.2007). The manganese content of the mushrooms studied, in the present work, was found to vary from 47.55 ± 1.78 mg/kg in *Pleurotus ostreatus* to 41.29 ± 1.77 mg/kg in *Pleurotus florida* (Table 2). The average manganese content of the analyzed samples was 44.42 mg/kg, which is far below the toxicity limit between 400 and 1000 mg/kg of manganese (WHO 1982). Usual manganese content in mushrooms varies between 10.0 and 60.0 mg/kg (Kalac 2010). Similar earlier studies have also indicated manganese contents in the mushrooms samples vary as follow: 29.5 – 46.8 mg/kg (Li et al. 2011), 7.2 – 55.6 mg/kg (Ouzouni et al. 2009), 37.9 – 38.7 mg/kg (Mleczek et al. 2013), 0.7 – 4.2 mg/kg (Radulescu et al. 2010), 2.2 – 48.5 mg/kg (George et al. 2014) and 1.3 – 47.5 mg/kg (Sen et al. 2012). Concentrations of manganese determined, in this study, are also in good agreement with the results reported in the literature.

Cadmium is accumulated mainly in kidneys, spleen and liver and its blood serum level increases considerably following mushroom consumption (Kalac and Svoboda 2000). Thus, cadmium seems to be the most deleterious among heavy metals in the mushrooms. Lead is similar to Cadmium having no beneficial effect in human metabolism, producing progressive toxicity. Lead can reach humans through air, water, and food. It accumulates in bones and can take in the role of calcium. Lead creates health disorders such as sleeplessness, tiredness, hearing, and weight loss (Udochukwu et al. 2014). Fortunately, the concentrations of the toxic heavy metals, cadmium and lead, in the studied mushroom samples were too low to be detected by the analytical technique used in this study. According to FAO/WHO (Kalac 2010) tolerable weekly intake of cadmium and lead are

0.007 and 0.025 mg/kg body weight, respectively. The lead and cadmium levels in all studied species are very low and thus, these mushroom species are safe for consumption against the risk of these metals.

The difference in values found in this study as compared to the values determined in other studies can be attributed to various factors such as the analytical methods used and the substrate on which they were cultivated. Mushrooms are known to undergo very effective mechanism that enables them to readily take up some heavy metals from the ecosystem (Zhu et al. 2011). The accumulation of heavy metals in the mushrooms seems to be affected by environmental and fungal related factors. Environmental factors such as organic matter content, pH and metal concentration in substrate and fungal factors such as the species, morphological part of the fruit body, developmental stages, age of the mycelium, intervals between fructifications and biochemical composition contribute significantly to the amount of heavy metals in the mushroom samples (Radulescu et al. 2010).

3.4 Metal contents in the substrates

As shown in Table 2, the mean concentration of copper in *Pleurotus florida* substrate was the highest (51.87 ± 2.06 mg/kg) and least in *Pleurotus ostreatus* (48.97 ± 3.14 mg/kg). The concentrations of copper found in the substrate samples for both mushroom samples are comparable. Similarly, comparable concentrations of iron were also found in the substrate of all study samples; 299.51 ± 11.55 mg/kg in*Pleurotus ostreatus* substrate and 265.23 ± 6.92 mg/kg for *Pleurotus florida* (Table 2). The concentration of zinc in the substrate samples is 62.55 ± 1.55 mg/kg and 64.25 ± 1.12 mg/kg for *Pleurotus ostreatus* and*Pleurotus florida,* respectively (Table 2). The concentration of manganese in the substrate sample is the highest in *Pleurotus florida* (62.55 ± 1.84 mg/kg) and the least in *Pleurotus ostreatus* (58.32 ± 1.98 mg/kg) (Table 2). On the other hand, concentrations of the toxic heavy metals, cadmium and lead, in the studied substrates were too low to be detected by the analytical technique used in the study. The mean concentration of metals in the substrate samples followed the decreasing order of: Fe > Zn > Mn > Cu > Cd and Pb both for *Pleurotus ostreatus* and *Pleurotus florida*.

In order to better understand the general characteristics of the substrates, in relation to the sites where the fruiting bodies were harvested, pH and humidity of the substrate have been determined. Accordingly, the humidity of the analyzed substrate samples possess relatively higher mean value, i.e. 74.80 %, which could be attributed to high moisture content of the analyzed substratum (wheat straw). The substrate pH was also found to be 6.87, which might probably be related to greater contents of biological materials. The mean content of trace metals, for example zinc, in the substrate was found to be comparable to the normal value detected in organic soil (57.0 – 100.0 mg/kg), and did not reached this limit for copper (1.0 – 115.0 mg/kg) (Elekes et al. 2010).

3.5 The bioaccumulation factor

The ability of mushrooms to accumulate elements from the substrates is expressed by bio-concentration factor (BCF), i.e., the ratio of the elemental content in the fruiting body to the content in underlying substrate (both in dry matter). In this context, more information is necessary from which substrate horizons the individual mushroom species takes their nutrients. It is usually the uppermost layer of 10 cm depth which is being sampled and analyzed for trace elements because this layer of the substrate contains mainly organic debris. However, mycelium of some mushroom species, e.g. *Suillus luteus*, is also associated with mineral horizons (Tedersoo et al. 2003). The coefficient of accumulation factor (BCF) of the heavy metals is calculated using the following relation:

$$\text{BCF} = \frac{\text{Concentration of heavy metal in Mushroom}}{\text{Concentration of heavy metal in Substrate}}$$

The accumulation factors of the target heavy metals in the mushroom samples were calculated as 1.047, 0.736, 1.434 and 0.815 in *Pleurotus ostreatus*; and 1.034, 0.920, 1.483 and 0.660 in *Pleurotus florida* for Cu, Fe, Zn and Mn, respectively. The corresponding results are shown in Table 5. Metal bio-accumulation factors in this study are comparable with those reported by Blanusa et al. (2001), Racz and Oldal (2000), Radulescu et al. (2010) and Stihi et al. (2011). While Cu and Zn being accumulated at higher ratios by the mushrooms, the other elements being accumulated at relatively lesser ratios.

Table 5. Mean bio-accumulation factors of heavy metals from substrate to mushrooms

Species	Cu	Fe	Zn	Mn
Pleurotus ostreatus	1.047	0.736	1.434	0.815
Pleurotus florida	1.034	0.920	1.483	0.660

3.6 Mushroom percentage contribution for daily needs in human body

Based on the mineral contents of the mushroom species, it is converted on the scale for 100 g fresh mushroom (based on the 90 % moisture content). This eventually signifies the minerals available in 100 g fresh mushrooms that can be consumed per day. The situation becomes quite practical when 100 g of fresh mushroom is consumed at least four or five times a week. Recommended dietary allowances (RDA) and tolerable upper intake levels (UL), that is likely to pose no risk of adverse health effects, of the metals along with relevant references are presented in Table 6.

Table 6. Recommended Daily Intake (RDI) and Tolerable Upper intake Limit (UL) of essential elements

Element	RDA	UL	Reference
Copper (Cu)	0.9 mg/day	10.0 mg/day	Ross et al. 2011
Iron (Fe)	8.0 – 18.0 mg/day	45.0 mg/day	Ross et al. 2011
Zinc (Zn)	8.0 – 11.0 mg/day	40.0 mg/day	Ross et al. 2011
Manganese (Mn)	1.8 –2.3 mg/day	11.0 mg/day	Ross et al. 2011

The concentrations of these essential metals reported to 100 g of fresh weight are calculated in Table 7, and in Table 8 also listed the percentage of these amounts in the daily physiological necessary for an adult. Based on these results, *Pleurotus florida* has the higher contribution of minerals in the daily necessary than *Pleurotus ostreatus* in 100 g of fresh mushrooms. The calculated percentage contributions of essential metals in this study are comparable with the data reported in Elekes and Dumitru (2010). Cu and Zn were accumulated at a relatively higher ratio by the mushroom samples while the other elements were being accumulated at a moderate ratio. Copper have greater contributions, 40.0 % to 53.3 %, which is also important for the daily nutrition of humans.

Table 7. Concentrations of the essential metals in the studied mushrooms samples (mg/100 g of fresh weight)

Species	Cu	Fe	Zn	Mn
Pleurotus ostreatus	0.36	1.57	0.64	0.34
Pleurotus florida	0.48	2.20	0.86	0.37

Table 8. The percentage contribution for the daily necessary in human body

Species	Cu	Fe	Zn	Mn
Pleurotus ostreatus	40.0 %	8.7 – 19.6 %	5.8 – 8.0 %	14.7 – 18.9 %
Pleurotus florida	53.3 %	12.2 – 27.5 %	7.8 – 10.7 %	16.1 – 20.5 %

3.7 Statistical analysis

In this study, the fruiting body of the mushroom and substrate samples were digested by three different digestion methods (dry, wet and microwave). Each sample was mixed thoroughly and one representative bulk sample was taken for each kind of samples. To this end, three aliquots were taken from each bulk sample and then digested and analyzed. During these processes, a number of random errors might be introduced in each aliquot and in each replicate measurement. There may be differences in results of the analysis between different mushroom samples within and between groups. Therefore, depending upon the type and the nature of results, there are different statistical methods used to check whether differences could occur between the results of analysis or not; and if there is a difference, statistical analysis signifies whether the difference is significant or not. Of these, one-way analysis of variance (ANOVA) has been selected for indicating the differences (Miller and Miller 2010).

Moreover, analysis of variance (ANOVA) is used to test hypothesis about differences between two or more means. For the present study, the significance of variation within three digestion methods has been calculated using one-way ANOVA. Microsoft Excel has also used to calculate the presence or absence of significant difference in mean concentration of each metal between different kinds of mushroom samples, substrate samples and digestion methods.

3.7.1 Statistical analysis of the mushroom samples

For *Pleurotus ostreatus* and *Pleurotus florida* mushroom samples, one way ANOVA has showed that except for copper ($p = 0.580$, $p = 0.185$) and manganese ($p = 0.314$, $p = 0.213$), respectively, the concentrations of iron and zinc have ($p < 0.05$) overall significant at 95 % confidence interval. The least significant difference (LSD) value was calculated for all metals and compared with the difference between the means of the metal contents between the three digestion methods. For zinc, all means differ significantly from each other while for copper and manganese, all the means were not differing significantly. For iron, all the means were found to differ significantly except the mean of the wet digestion which was not differing significantly from microwave digestion.

Obviously, presence of the significant difference between the means may be associated with the different digestion methods employed for sample preparation. The differences in the contents of some of the metals have been reported to be heavily dependent upon the composition of the substrate from which the mushrooms got their nutrients, the site where the samples collected from, age of the fruiting bodies and mycelium, and distance from the source of pollution (Kalac et al. 1991; Radulescu et al. 2010). Similarly, presence of significant difference in concentration for some minerals may indicate that either the studied samples contain higher concentration of mineral nutrients in the substrate or a well-aged mushroom has been harvested and good precautions have been followed during processing.

3.7.2 Statistical analysis of the substrate samples

For *Pleurotus ostreatus* substrate sample, a one way ANOVA showed that except for copper ($p = 0.06$), the concentrations of iron, zinc and manganese were ($p < 0.05$) overall significant at 95 % confidence interval. The least significant difference (LSD) value was also calculated for all metals and compared with the difference between the means of the metal contents between the three digestion methods. For iron, all means differ significantly from each other while for copper all means don't differ significantly except for dry ashing with microwave digestion. For zinc and manganese, all means differ significantly except for the mean of the wet digestion which was not found to differ significantly from the microwave digestion.

For *Pleurotus florida* substrate sample, a one way ANOVA shows that except for copper ($p = 0.08$), the concentrations of iron, zinc and manganese were ($p < 0.05$) overall significant at 95 % confidence interval. The least significant difference (LSD) values were also calculated for all metals and compared with the differences between the means of the metal contents between the three digestion methods. For copper, iron, zinc and manganese, all the means were found to differ significantly from each other except for the mean of the wet digestion which was not noted to differ significantly from the microwave digestion.

3.8 Comparison of metal content of the current study with earlier reported results

There are several studies carried out on analysis of the metal contents of different mushroom species, in different countries of the world. Therefore, it is reasonable to compare the results obtained from the analysis of the mushroom species considered in this study with the values reported, in the literature, for other mushroom species analyzed elsewhere. The outcome of this comparison may help to identify the possible differences in composition and if there exists a deviation from certain guide lines. To this end, comparison of metal concentrations in the Ethiopian mushroom species with those reported from other countries has been done and given in Table 9.

Table 9. Comparison of metal concentration (mg/kg dry weight) of different mushroom species from different countries

Country	Species	Concentration (mg/kg dry weight) of metals in mushrooms						References
		Cu	Fe	Zn	Mn	Cd	Pb	
Turkey	*Pleurotus sajur-caju*	8.0	533.0	35.0	5.1	ND	ND	Akyuz and Kirbag 2010
Poland	*Rozites caperatus*	66.3	329.9	152.5	38.3	3.98	1.20	Mleczek et al. 2013
China	*Agaricus bisporus*	14.8	190.0	81.4	28.8	0.35	2.2	Zhu et al. 2011
India	*Agaricus bisporus*	7.4	1596.0	46.8	16.7	0.14	0.75	Naresh et al. 2012
Nigeria	*Polyporus frondosis*	19.45	-	38.05	18.04	0.47	0.95	Ita et al. 2008
	Plerotussquarrosulus	1.75	33.2	16.9	-	3.56	1.56	Udochukwu et al. 2014
USA	*Lentinula edodes*	10.9	36.7	34.4	25.3	1.98	1.26	George et al. 2014
Romania	*Lycoperdon Perlatum*	11.8	623.0	127.0	12.6	1.64	0.71	Stihi et al. 2011
	Agaricus campestris	10.0	261.7	96.6	2.4	0.19	1.56	Radulescu et al. 2010
Greece	*Lepista nuda*	75.06	74.6	98.99	33.65	0.25	1.16	Ouzouni et al. 2009
Ethiopia	*Pleurotus ostreatus*	51.19	220.87	89.68	47.55	ND	ND	This study
Ethiopia	*Pleurotus florida*	53.56	243.92	95.26	41.29	ND	ND	This study

The levels of copper in the currently analyzed species ranged from 51.19 to 53.56 mg/kg. The observed copper concentrations were higher than those reported from other countries for different mushroom species except for the result reported by Ouzouni et al. 2009 from Greece. The amount of iron determined in this study was higher than those reported from China (Zhu et al. 2011), Greece (Ouzouni et al. 2009), USA (George et al. 2014) and Nigeria (Udochukwu et al. 2014). However, the levels of iron found in the presently analyzed species were comparable with those reported from Romania (Stihi et al. 2011; Radulescu et al. 2010), Poland (Mleczek et al. 2013) and Turkey (Akyuz and Kirbag 2010) for other mushroom species. Zinc is present in appreciable amount in the mushroom species analyzed. The observed concentrations of zinc were higher than those mushroom species from China (Zhu et al. 2011), Nigeria (Ita et al. 2008; Udochukwu et al. 2014), India (Naresh et al. 2012), USA (George et al. 2014) and Turkey (Akyuz and Kirbag 2010). The zinc concentrations in the studied samples were found to be in agreement with those obtained from Romania (Stihi et al. 2011; Radulescu et al. 2010) and Greece (Ouzouni et al. 2009). The amount of manganese determined in the current study was also in good agreement with the ones reported from different countries for different mushroom species; corresponding results given in Table 9. Regarding the non-essential metals (Cd and Pb), in most mushroom species, the amounts of cadmium and lead were detected at very low concentration levels. Similarly, in the present study, the levels of these two metals were found to be below their detection limits.

As has also been indicated above, the results obtained in the presented study were noted to be favorably comparable with the reported findings. However, results for some metals from mushroom species considered, in this study, were observed to differ greatly. This difference in the contents of the metals could be associated to the variation in the composition of the substrate from which the mushrooms got their nutrients, the site where the samples collected from, age of the fruiting bodies and mycelium, and distance from the source of pollution (Kalac et al. 1991; Radulescu et al. 2010). Besides, the metal concentrations in the mushroom samples could also mainly be affected by the pH or organic matter content of the substrates (Gast et al. 1988; Demirbas 2002).

4. CONCLUSIONS

The levels of essential and non-essential metals, *viz.*, Cu, Fe, Zn, Mn, Cd and Pb in the fruiting bodies of the two oyster mushroom species (*Pleurotus ostreatus* and *Pleurotus florida*) and their respective substrate samples have been analyzed utilizing flame atomic absorption spectrometry after digesting the samples employing dry ashing, wet digestion and microwave digestion techniques. The use of microwave digestion system provides very fast, safer, simpler and cleaner sample preparation method than both dry ashing and wet digestion method.

The levels of metals analyzed in the samples were also compared with those reported in the literature. It was observed, in the present study, that the detected levels of copper, zinc, iron, and manganese were generally in agreement with the reported results. The uptake of metals by the mushroom samples may be different from one species to others and their respective substrates, in many ways. Mushrooms accumulate Zn and Cu to higher extents than Fe and Mn. Therefore, determination of heavy metal concentrations in the fruiting bodies of oyster mushrooms is essential in dietary intake studies, because mushrooms constitute important parts of the diet. In this regards, the two mushroom species considered in this study and available in the local markets in Ethiopia, could be recommended for consumption from the point of view of accumulation of the heavy metals analyzed.

The elemental concentrations of the mushrooms in the present study are approximately of similar to those reported in the literature. It should also be noted that some discrepancy is possible due to the different analytical techniques used in other studies on mushrooms. The uptake of different forms and complexes of elements in mushrooms should, however, be considered in more detail as the impacts on human health may vary depending on the chemical forms of the element ingested. Further investigations on the mechanisms of possible accumulations in various tissues and also the synergistic toxic effects of multiple elements will be the aspects of considerable attentions of the authors.

COMPETING INTERESTS

The authors declare that they have no competing interests.

AUTHORS' CONTRIBUTIONS

MG contributed to the sampling and analysis. MG, NM and AMT contributed to the design of the study. MG and NM contributed to the interpretation of data and drafting the manuscript. NM contributed to critical revision of the paper. All authors read and approved the final manuscript.

REFERENCES

1. Agrahar-Murugkar D, Subbuakshmi G. Nutritional value of edible wild mushrooms collected from the Khasi Hills of Meghalaya. Food Chem. 2005;89:599–603.

2. Akyuz M, Kirbag S. Nutritive value of wild edible and cultured mushrooms. Tur J Bio. 2010;34:97–102.

3. Alemaya University of Agriculture. Proceedings of the 13th annual research and extension review meeting, Alemaya University of Agriculture, Alemaya, Ethiopia, 26–28 February 1996. 1996.

4. Baslar S, Kula I, Dogan Y, Yildiz D, Ay G. A study of trace element contents in plants growing at Honaz Dagi-Denizli, Turkey. Eko. 2009;18:1–7.

5. Blanusa M, Kucak A, Varnai VM, Saric MM. Uptake of cadmium, copper, iron, manganese and zinc in mushrooms (*Boletaceae*) from Croatian forest soil. J AOAC Inter. 2001;84:1964–71.

6. Demirbas A. Heavy metal bioaccumulation by mushrooms from artificially fortified soils. Food Chem. 2001;74:293–301.

7. Demirbas A. Metal ion uptake by mushrooms from natural and artificially enriched soils. Food Chem. 2002;78:89–93.

8. Diez VA, Alvarez A. Compositional and nutritional studies on two wild edible mushrooms from Northwest Spain. Food Chem. 2001;75:417–22.

9. Dulay RMR, Castro MEGD, Coloma NB, Bernardo AP, Cruz AGD, Tiniola RC, Kalaw SP, Reyes RG. Effects and myco-accumulation of lead (Pb) in five pleurotus mushrooms. Inter J Bio Phar All Scie. 2015;4(3):1664–77.

10. Elekes CC, Busuioc G, Ionita G. The bioaccumulation of some heavy metals in the fruiting body of wild growing mushrooms. Not Bot Hort Agrobot Cluj-Napo. 2010a;38(2):147–51.

11. Elekes CC, Dumitru M. Trace metals concentrations in some wild grown edible mushrooms from wooded area of Romania. Paper presented at the WSEAS International Conference on Environment, Medicine and Health Sciences, At Penang, Malaysia, 23–25 March 2010. 2010.

12. Eurachem/Co-Operation on International Traceability in Analytical Chemistry Guide. Traceability in Chemical Measurement, https://www.eurachem.org/images/stories/Guides/pdf/EC_Trace_2003.pdf. Accessed 12 March 2015 2003.

13. European Commission. Document on pesticide residue analytical methods. http://ec.europa.eu/food/plant/pesticides/guidance_documents/docs/guide_doc_825-00_rev8_en.pdf. Accessed 09 April 2015. 2010

14. Garcia-Lafuente A, Moro C, Villares A, Guillamón E, Rostagno MA, D'Arrigo M, Martínez JA. Mushrooms as a source of anti-inflammatory agents. Am J Comm Psychol. 2011;48:125–41.

15. Gast GH, Jansen E, Bierling J, Haanstra L. Heavy metals in mushroom and their relationship with soil characteristics. Chemo. 1988;60:789–99.

16. Genccelep H, Uzun Y, Tuncturk Y, Demirel K. Determination of mineral contents of wild-grown edible mushrooms. Food Chem. 2009;113:1033–6.

17. George PL, Ranatunga TD, Reddy SS, Sharma GC. A comparative analysis of mineral elements in the mycelia and the fruiting bodies of Shiitake mushrooms. Am J Food Tech. 2014;9(7):360–9.

18. Isiloglu M, Yilmaz F, Merdivan D. Concentrations of trace elements in wild edible mushrooms. Food Chem. 2001;73:169–75.

19. Ita BN, Ebong GA, Essien JP, Eduok SI. Bioaccumulation potential of heavy metals in edible fungal *Sporocarps*from the Niger Delta region of Nigeria. Pak J Nutr. 2008;7:93–7.

20. Kalac P. Trace element contents in European species of wild growing edible mushrooms: A review for the period 2000–2009. Food Chem. 2010;122:2–15.

21. Kalac P, Svoboda L. A review of trace element concentrations in edible mushrooms. Food Chem. 2000;69:273–81.

22. Kalac P, Burda J, Staskova I. Concentrations of lead, cadmium, mercury, and copper in mushrooms in the vicinity of a lead smelter. Scie Tot Env. 1991;105:109–19.

23. Kues U, Liu Y. Fruit body production in basidiomycetes. App Micro Biotech. 2000;54:141–52.

24. Li T, Wang Y, Zhang J, Zhao Y, Liu H. Trace element content of Boletus tomentipes mushroom collected from Yunnan, China. Food Chem. 2011;127:1828–30.

25. Manzi P, Aguzzi A, Pizzoferrato L. Nutritional value of mushrooms widely consumed in Italy. Food Chem. 2001;73:321–5.

26. Miller JN, Miller JC. Statistics and Chemometrics for Analytical Chemistry. England: Pearson Harlow; 2010.

27. Mleczek M, Magdziak Z, Goliński P, Siwulski M, Stuper-Szablewska K. Concentrations of minerals in selected edible mushroom species growing in Poland and their effect on human health. Acta Sci Pol Tech Aliment. 2013;12(2):203–14.

28. Naresh RM, Udaya BRE, Byragi RT. Assessment of heavy metal threat in agaricus bisporus mushrooms cultivated from water hyacinth weed of Kolleru lake, Andhra pradesh-India. Int J Env Sci. 2012;3(1):28–35.

29. Obodai M, Ferreira ICFR, Fernandes A, Barros L, Mensah DLN, Dzomeku M, Urben AF, Prempeh J, Takli RK. Evaluation of the chemical and antioxidant properties of wild and cultivated mushrooms of Ghana. Mole. 2014;19:19532–48.

30. Okwulehie IC, Ogoke JA. Bioactive, nutritional and heavy metal constituents of some edible mushrooms found in Abia State of Nigeria. Int J App Microbio Biotech Res. 2013;1:7–15.

31. Ouzouni PK, Petridis D, Koller WD, Riganakos KA. Nutritional value and metal content of wild edible mushrooms collected from West Macedonia and Epirus, Greece. Food Chem. 2009;115:1575–80.

32. Quarcoo A, Adotey G. Determination of heavy metals in *Pleurotus ostreatus* (Oyster mushroom) and*Termitomyces clypeatus* (Termite mushroom) sold on selected markets in Accra, Ghana. Mycosph. 2013;4(5):960–7.

33. Racz L, Oldal V. Investigation of uptake processes in a soil/mushroom system by AES and AAS methods. Microchem J. 2000;67(1–3):115–8.

34. Radulescu C, Stihi C, Busuioc G, Popescu IV, Gheboianu AI, Cimpoca VG. Evaluation of essential elements and heavy metal levels in fruiting bodies of wild mushrooms and their substrate by EDXRF spectrometry and FAA spectrometry. Rom Biotech Lett. 2010;15(4):5444–56.

35. Ragunathan R, Swaminathan K. Nutritional status of *Pleurotus spp.* grown on various agro-wastes. Food Chem. 2003;80:371–5.

36. Ross AC, Taylor CL, Yaktine AL, Del Valle HB, Committee to Review Dietary Reference Intakes for Vitamin D and Calcium; Food and Nutrition Board; Institute of Medicine. Dietary Reference Intakes for Calcium and Vitamin D. Washington D.C: National Academy Press; 2011.

37. Sara M. Effect of spawn substrates and their inoculation rates on the yield and quality of oyster mushroom (*Pleurotus florida*). M.Sc. Thesis, Haramaya University. 2007.

38. Sarikurkcu C, Copur M, Yildiz D, Akata I. Metal concentration of wild edible mushrooms in Soguksu National Park in Turkey. Food Chem. 2011;128:731–4.

39. Schillaci D, Arizza V, Gargano ML, Venturella G. antibacterial activity of mediterranean oyster mushrooms, species of genus pleurotus (higher basidiomycetes). Int J Med Mushroom. 2013;15:591–4.

40. Sen I, Alli H, Col B, Celikkollu M, Balci A. Trace metal contents of some wild-growing mushrooms in Bigadiç (Balıkesir), Turkey. Turk J Bot. 2012;36:519–28.

41. Sesli E, Tuzen M, Soylak M. Evaluation of trace metal contents of some wild edible mushrooms from Black Sea region, Turkey. J Haz Mat. 2008;160:462–7.

42. Shukla CS, Biswas MK. Evaluation of different techniques for oyster mushroom cultivation. J Myco Pla Patho. 2000;30:431–5.

43. Silvestre MD, Lagarda MJ, Farra R, Martineze-Costa C, Brines J. Copper, iron and zinc determination in human milk using FAAS with microwave digestion. Food Chem. 2000;68:95–9.

44. Singh RP, Chaube HS. Mushroom production technologies: mushroom research laboratory. Pant University of Agriculture and Technology. India: Pantnagar; 1995.

45. Smith JF, Margarel E. Investigation on to the cultural requirement of a brown capped *Agaricus* strain (W4II) isolated from *curpressus* leaf litter. J Hortic Sci. 1995;70:963–74.

46. Soylak M, Saracoglu S, Tuzen M, Mendil D. Determination of trace metals in mushroom sample from Kayseri, Turkey. Food Chem. 2005;92:649–52.

47. Stihi C, Radulescu C, Busuioc G, Popescu IV, Gheboianu A, Ene A. Studies on accumulation of heavy metals from substrate to edible wild mushrooms. Rom J Phy. 2011;56(1–2):257–64.

48. Svoboda L, Zimmermannova K, Kalac P. Concentrations of mercury, cadmium, lead and copper in fruiting bodies of edible mushrooms in an emission area of a copper smelter and a mercury smelter. Scie Tot Env. 2000;246:61–7.

49. Tedersoo L, Koljalg U, Hallenberg N, Larsson KH. Fine scale distribution of ectomycorrhizal fungi and roots across substrate layers including coarse woody debris in a mixed forest. New Phyto. 2003;159:153–65.

50. Tuzen M, Sesli E, Soylak M. Trace element levels of mushroom species from East Black Sea region of Turkey. Food Cont. 2007;18:806–10.

51. Udochukwu U, Nekpen BO, Udinyiwe OC, Omeje FI. Bioaccumulation of Heavy metals and pollutants by edible mushroom collected from Iselu market Benin-city. Int J Curr Microbiol App Sci. 2014;3(10):52–7.

52. Unak P, Lambrecht FY, Biber FZ, Darcan S. Iodine measurements by isotope dilution analysis in drinking water in Western Turkey. J Radioanal Nucl Chem. 2007;273:649–51.

53. Uzun Y, Genccelep H, Kaya A, Akcay ME. The mineral contents of some wild edible mushrooms. Eko. 2011;20(80):6–12.

54. World Health Organization. Evaluation of Certain Foods Additives and Contaminants (Tweenty-Six Report of the Joint FAO/WHO Expert Committee on Food Additives). WHO Technical Report series, No. 683 Geneva. http://apps.who.int/iris/bitstream/10665/41546/1/WHO_TRS_683.pdf. Accessed 05 December 2015. 1982.

55. Yoshida S, Muramatsu Y. Determination of major and trace elements in mushroom, plant and soil samples collected from Japanese forests. Int J Env Ana Chem. 1997;67:49–58.

56. Zadrazil F, Dube HC. The oyster mushroom. Importance and prospects. Mush Resea. 1992;1(1):25–32.

57. Zhu F, Qu L, Fan W, Qiao M, Hao H, Wang X. Assessment of heavy metals in some wild edible mushrooms collected from Yunnan Province, China. Env Mon Asse. 2011;179:191–9.

Chapter 6

Levels of major and trace elements in fennel (*Foeniculum vulgari* Mill.) fruits cultivated in Ethiopia

Feleke Demissie Endalamaw[1,2] and Bhagwan Singh Chandravanshi[1*]

[1]Department of Chemistry, College of Natural Sciences, Addis Ababa University, P.O. Box 1176, Addis Ababa, Ethiopia.

[2]Department of Chemistry, Faculty of Natural and Computational Sciences, Wolayta Sodo University, P.O. Box: 138, Wolayta Sodo, Ethiopia.

ABSTRACT

Background

Sweet fennel (*Foeniculum vulgare* Mill.) is one of the precious spices. Almost all parts of fennel plant are edible. The herb is used as carminative, digestive, diuretic, cosmetic and medicine.

Methods

A 0.5 g of the oven dried fennel fruit and soil samples were digested by wet-digestion method. The levels of selected elements (Ca, Mg, Fe, Mn, Cu, Cr, Co, Zn, Ni, Cd and Pb) were determined in sweet fennel fruit

(*Foeniculum vulgari* Mill.) and soil from Addis Ababa (Central Ethiopia) and Gojjam (Northern West Ethiopia) by flame atomic absorption spectrometry (FAAS).

Results

The elemental concentrations (µg/g) in fennel fruit were: Ca (20,500–23,000), Mg (1,310–3,460), Fe (1,140–1,900), Mn (31–51), Cu (24–103), Cr (91–98), Co (26–71), Zn (37–45), Ni (19–24), and Cd (1.6–1.9) while in the soil were: Ca (1,440–1,780), Mg (1,260–3,310), Fe (26,900–28,000), Mn (1,460–1,980), Cu (51–101), Cr (127–141), Co (54–143), Zn (99–104), Ni (98–161), and Cd (1.7–2.9). Pb was below the method detection limit in both the fennel fruit and soil.

Conclusion

The Ethiopian fennel fruits are rich in Ca and Mg and other essential elements (Fe, Cu, Co and Zn) and can be used as good supplement for human being in particularly for children and pregnant women. The toxic element Cd is at trace level and Pb is not detected in the fennel fruit. Thus, Ethiopian fennel fruits are safe for human consumption.

KEYWORDS

Fennel Foeniculum vulgari Mill, Ensilal Trace elements, Spice Ethiopia

1. INTRODUCTION

Sweet fennel (*Foeniculum vulgare* Mill.) is one of the precious spices. It is a highly aromatic (Parthasarathy et al. 2008) and flavourful herb with culinary, cosmetic (Rasul et al. 2012; Kim et al. 2004) and medicinal uses (Javidnia et al. 2003; He and Huang 2011; Cetin et al. 2010; Faudale et al. 2008; Mohamad et al. 2011; Qiu et al. 2012; Senatore et al. 2013; Shahat et al. 2011; Dadalioglu and Evrendilek2004). It has also nutritional values

(Renna and Gonnella 2012). It belongs to the family Apiaceae (formerly the Umbelliferae) (Qiu et al. 2012). In Ethiopia, the plant is called *Ensilal*. It is native to Southern Europe and the Mediterranean region and is cultivated mainly in India, China, Guatemala, Sri Lanka, Greece, and many other countries in Europe and America (Parthasarathy et al. 2008; He and Huang2011; Rather et al. 2012; Rahimi and Ardekani 2013; Chéry et al. 2008). As to our knowledge, it is cultivated only in Ethiopia in Africa.

Almost all parts of fennel plant are edible. The herb is used as carminative, digestive, and diuretic. It is used in treating respiratory and gastrointestinal disorders (Diao et al. 2014; El-Motaium and El-Seoud2007). Furthermore, its insect rippling and antimicrobial properties against a wide range of microorganisms have been well established (Qiu et al. 2012; Oktay et al. 2003). Fennel essential oil is used in cosmetics, pharmaceuticals, and perfumery and as a food additive and flavouring agent in food products (Kim et al. 2004; Mohamad et al. 2011; Rather et al. 2012; Oktay et al. 2003). The fennel fruits are used as flavourings in baked goods, meat and fish dishes, ice cream, alcoholic beverages and herb mixtures. Fennel fruit tea and fennel fruit water are very common traditional medicines for flatulence treatment (He and Huang 2011; Rather et al. 2012). In China, fennel fruit is used both as tradition Chinese medicine and as common food with Chinese name *Xiaohuixiang* (He and Huang 2011).

There are various demands for elemental information in biological and medical samples. Trace elements exhibit significant functions with respect to living organisms in terms of oxygen carrying, electron transfer, enzyme catalyzing, etc. (Singh and Garg 2006; Divrikli et al. 2006; Demirel et al. 2008). For all the elements essential for metabolism, there exists a range of intake over which their supply is adequate for the body beyond which deficiency and toxic effects are observed (Mallinckrodt and Meissne 1994). All the nutrient elements are primarily supplied through diet. However, this may change, depending on age, sex, health status, geographical and climatic conditions (O'Dell and Sunde 1997). Therefore, it is essential to determine elemental contents of food items. WHO and FAO have recommended selective studies of individual foodstuffs as an important step in the estimation of dietary intake of trace elements (FAO/WHO2001). It is a good practice to include micronutrient density of the diet component in addition to energy and protein adequacy of it (FAO/WHO 2001).

There are some works done on the elemental level of infusion and decoction tea (Desideri et al. 2011; Ozcan et al. 2008; Szymczycha-Madeja et al. 2012; Sembratowicz and Rusinek-prystupa 2014; Başgel and

Erdemoğlu 2006; Zengin et al. 2008) of the fennel fruit, steam and leaf. Fennel fruit is eaten cooked or raw in different part of the world. Chewing fennel fruit after a meal is a common practice in India (Shirahatt et al. 2010). Ethiopians use raw or roasted and ground form of it as spice. It is also used to flavor local drinks such as *areke* and *tej*. Moreover, in Ethiopia the fennel fruit is used as a spice in the preparation of *shiro* and *berbere* and the powdered roasted fennel fruit is added in to traditional dish like*doro wot* (Getahun 1976; Zuberi et al. 2014). There is no report on the elemental content of fennel fruit in Ethiopia in any form and only few reports (Khattak and Khattak 2011; Chowdhary et al. 2014; Ozcan and Akbulut 2007; Kumar et al. 2005) are available worldwide on the total elemental content of the fennel fruit. None of them correlate the nutrient found in the fennel fruit with the soil it is grown. This study was therefore; designed to assess the levels of major and trace elements (Ca, Mg, Fe, Mn, Cu, Cr, Co, Zn, Ni, Cd and Pb) in fennel fruit cultivated in the farmland in Ethiopia. To assess the significance of soil to the growth of fennel, it was also analysed for major and trace elements using flame atomic absorption spectrometry.

2. METHODS

2.1 Instruments and reagents

Blending device (Moulinex, France), digital analytical balance (Adam, Model WL 3000, Switzerland), hot plate, air-circulating oven (Digitheat, J.P. Selecta, Spain) and flame atomic absorption spectrophotometer (FAAS) (Buck Scientific, Model 210VGP AAS, East Norwalk, USA) were used.

Chemicals and reagents that were used in the analysis were all analytical grades. 69–72% HNO_3(Supreme Enterprises Cantt, India), 70% $HClO_4$ (A.C.S. Reagent, Aldrich, UK) and 36% H_2O_2 (Scharlau Chemie, European Union, UN2014) were used for digestion of powdered fennel fruit and soil samples. Lanthanum nitrate hydrate (99.9%, Aldrich, Muwaukee, USA) was used to prevent the chemical interference of phosphate ion on Ca and Mg during the analysis of the samples. Stock standard solution of concentration 1000 mg/L in 2% HNO_3 of the elements Ca, Mg, Mn, Fe, Cu, Zn, Ni, Co, Cr, Pb and Cd (Buck Scientific Puro-Graphic) standard solutions were used to prepare intermediate standard solutions to obtain calibration curves for the determination of elements in the samples and recovery study.

Table 1. Latitude, longitude, altitude, annual average temperature and annual rain fall of the sites

Name of sites	Latitude (N)	Longitude (E)	Altitude (m)	Annual average temperature (°C)	Annual average rain fall (mm)
AA	8°59′	38°48′	2,354	12-20	1,210
B	10°27′	38°12′	2,539	9.8-22	1,250
DM	10°21′	37°43′	2,440	14.2-18.6	1,340
FS	10°42′	37°16′	1,905	9.8-23.5	1,450

AA = Addis Ababa, B = Bichena, DD = Debre Markos and FS = Finote Selam.

2.2 Sampling and sample preparation

The study locations were Addis Ababa, Bichena, Debre Markos and Finote Selam. Addis Ababa is the capital city of Ethiopia. Bichena and Debre Markos are located in east Gojjam (Ethiopia); they are 299 km and 265 km from Addis Ababa, respectively. Finote Selam is found in west Gojjam, which is 380 km away from the capital city. Addis Ababa was selected because it is expected to be relatively contaminated area and to see the accumulation effect of the plant. Gojjam was selected because fennel is consumed in these zones every day in different ways, as spice, flavouring agent in local *areki* and used as medicinal plant. Latitude, longitude, altitude, annual average temperature and rainfall of the sampling sites of fennel fruit are given in Table 1.

Fennel fruit samples were collected during the dry season (March – April 2010). A total of 5 sub-sites were selected for each site and 100 g fennel fruit and 100 g soil were collected from each sub-site. The sub-sites were selected randomly. The samples collected in each sub-sites were pooled and 0.5 kg of bulk sample were collected for each of the sites. The soil samples were collected at a depth of 20 cm from the surface of the top soil. Samples were sealed in polyethylene bags and transported to the laboratory. All the samples were dried in the oven at 70°C for 48 hours for fennel fruit and 72 hours for soil until the mass of the samples became constant.

2.3 Optimization of digestion procedure

In this study nitric and perchloric acids were used during optimization procedure for fennel fruit and a total volume of acid was taken to be 4 mL. The ratio of the acids, temperature and time of digestion were optimized. The results are given in Table 2. Therefore, the procedure with total of 4 mL reagents volume (2.5 mL HNO_3 and 1.5 mL $HClO_4$), heating at 210°C and 150 min digestion time was selected for this study. Similar optimization also carried out for the soil sample digestion using HNO_3, $HClO_4$ and H_2O_2. The results are given in Table 3. The optimum condition obtained was 2 mL HNO_3 (69–72%), 4 mL $HClO_4$(70%) and 1.5 mL of H_2O_2 (36%) and at temperature 270°C for 180 min. The solid dry sample (for both soil and fennel fruit) digested by this procedure was 0.5 g.

Fennel fruit samples were digested by applying the optimized procedure. A 0.5 g of well-powdered fennel fruit sample was added into a round bottom flask (100 mL). To this flask 2.5 mL HNO_3 (69–72%) and 1.5 mL $HClO_4$ (70%)

were added and the mixtures were digested on a micro Kjeldahl digestion apparatus by setting the temperature at 210°C for 150 min.

Table 2. Reagent types and volumes, temperature and time attempted during optimization of digestion of 0.5 g fennel fruit

No.	Reagent volume (mL)			Temperature (°C)	Time (min)	Results
	HNO_3	$HClO_4$	Total			
1	1	3	4	270	180	Yellow
2	1.5	2.5	4	270	180	Lightly yellow
3	2	2	4	270	180	Almost clear
4	**2.5**	**1.5**	**4**	**270**	**180**	**Clear solution**
5	3	1	4	270	180	Clear solution
6	3.5	0.5	4	270	180	Almost clear
7	2.5	1.5	4	120	180	Yellow
8	2.5	1.5	4	150	180	Lightly yellow
9	2.5	1.5	4	180	180	Almost clear
10	**2.5**	**1.5**	**4**	**210**	**180**	**Clear solution**
11	2.5	1.5	4	240	180	Clear solution
12	2.5	1.5	4	270	180	Clear solution
13	2.5	1.5	4	270	105	Light brown with suspension
14	2.5	1.5	4	270	120	Yellow
15	2.5	1.5	4	270	135	Lightly yellow
16	**2.5**	**1.5**	**4**	**270**	**150**	**Clear solution**
17	2.5	1.5	4	270	165	Clear solution
18	2.5	1.5	4	270	180	Clear solution

Rows with bold font indicate the optimal condition for the given parameter.

Table 3. Reagent types and volumes, temperature and time attempted during optimization of digestion of 0.5 g soil

No.	Reagent volume (mL)			Temperature (°C)	Time (min)	Results
	HNO_3	$HClO_4$	H_2O_2			
1	2.2	4.4	1	300	180	Deep yellow with suspension
2	**2**	**4**	**1.5**	**300**	**180**	**Light yellow with no suspension**
3	1.8	3.6	2	300	180	Light yellow with no suspension
4	2	4	1.5	240	180	Deep yellow with suspension
5	**2**	**4**	**1.5**	**270**	**180**	**Light yellow with no suspension**
6	2	4	1.5	300	180	Light yellow with no suspension
7	2	4	1.5	270	120	Deep Yellow with suspension
8	2	4	1.5	270	150	Light yellow with suspension
9	**2**	**4**	**1.5**	**270**	**180**	**Light yellow with no suspension**

Rows with bold font indicate the optimal condition for the given parameter.

Soil samples were digested by applying the optimized procedure. A 0.5 g of well-powdered soil sample was added into a round bottom flask (100 mL). To this flask 2 mL HNO_3 (69–72%), 4 mL $HClO_4$ (70%) and 1.5 mL of H_2O_2 (36%) were added and the mixtures were digested on a micro Kjeldahl digestion apparatus by setting the temperature at 270°C for 180 min.

After cooling of the digest, about 15 mL of de-ionized water was added to dissolve the precipitates formed on cooling and to dilute left over acid so that to minimize the dissolution of a filter paper by the digest residue while filtrating with Whatman, (110 mm, dia), filter paper. The round bottom flask was rinsed by using 8 mL de-ionized water two times and filtered. The filtrate was filled to 50 mL with de-ionized water. Triplicate digestions were carried out for each sample of soil and fennel fruit. Six blank solutions were prepared by digesting the mixture of reagents following the same procedure. The digested samples were kept in the refrigerator, until the level of all the elements in the sample solutions were determined by FAAS.

2.4 Instrument calibration

Six points calibration curve were established by running series of the prepared working standard solutions. Correlation coefficients obtained for calibration curve were > 0.999. Immediately after calibration, the sample solutions were aspirated into the FAAS instrument and direct readings of the elemental concentrations was recorded. Calibration of the instrument was checked periodically during operation. Three replicate determinations were carried out on each sample. Blank analysis was also done similarly.

2.5 Method validation

The method detection limit of each element was calculated as tree times the standard deviation of the blank ($3\sigma_{blank}$, n = 6). The method detection limit for fennel fruit and soil samples are presented in Table 4. The method detection limits are low enough which clearly indicate that the selected elements can be detected at trace level by the FAAS method used.

Table 4. Method detection limits for elements (for all metals, n = 6)

Element	Ca	Mg	Mn	Fe	Zn	Cu	Cr	Co	Ni	Cd	Pb
MDL for fennel fruit (μg/g)	3	0.4	0.5	0.2	0.6	0.5	3	4	1	0.1	5
MDL for soil (μg/g)	5	0.5	1.5	0.4	2	2	3	3	3	1	5

MDL = Method detection limit.

To check the accuracy of digestion procedure and efficiency of the FAAS instrument spiked samples were prepared by adding a small known quantity of elemental standard solutions to both the soil and fennel fruit by applying similar digestion procedure, analysing for the levels of elements and calculating the recovery percent. The obtained percentage recovery for the fennel fruit varied from 93% to 108% and for the soil from 93% to 110% which are in the acceptable range. The reproducibility of the analytical procedure was checked by carrying out a triplicate analysis and calculating the coefficient of variation of the mean for each metal. In almost all cases, triplicate results did not differ by more than 10% of the mean.

2.6 Statistical analysis

The multivariate analysis of variance for the equality of means and correlation between the elements in fennel fruit and soil samples were done using the SPSS 16.0.

3. RESULTS

The fennel fruit and the soil samples on which the plant was grown were analyzed for major (Ca, Mg, Fe, and Mn), trace essential (Cu, Zn, Cr and Co), and trace non-essential elements (Ni, Cd and Pb) with FAAS. The mean concentrations along with standard deviation of triplicate analysis are given from Tables 5 and 6 for the fennel fruit and soil samples, respectively.

Table 5. Average concentration of elements (μg/g dry weight) in the fennel fruit samples from the four sites (mean ± SD, n = 3)

Element	Ca	Mg	Fe	Mn	Cu	Cr	Co	Zn	Ni	Cd	Pb
AA	20,500 ± 290	2,390 ± 68	1,140 ± 67	45.8 ± 1.1	103 ± 4	97.7 ± 2.9	67.7 ± 3.7	37.1 ± 2.0	18.7 ± 0.5	1.90 ± 0.10	ND
B	23,000 ± 240	1,970 ± 83	1,900 ± 81	51.4 ± 1.3	48.4 ± 0.6	91.4 ± 2.4	26.2 ± 0.3	37.8 ± 2.2	24.2 ± 1.1	1.67 ± 0.10	ND
DD	21,400 ± 100	3,460 ± 93	1,480 ± 62	48.1 ± 6.2	62.5 ± 0.3	94.0 ± 4.7	49.5 ± 0.9	44.7 ± 2.5	24.0 ± 0.7	1.89 ± 0.08	ND
FS	22,900 ± 570	1,310 ± 58	1,470 ± 74	30.6 ± 0.7	23.9 ± 0.2	90.9 ± 0.4	70.8 ± 0.6	40.8 ± 1.3	24.2 ± 0.6	1.59 ± 0.10	ND

AA = Addis Ababa, B = Bichena, DD = Debre Markos, FS = Finote Selam and ND = not detected.

Table 6. Average concentration of elements (µg/g dry weight) in the soil of the four sites (mean ± SD, n = 3)

Element	Ca	Mg	Fe	Mn	Cu	Cr	Co	Zn	Ni	Cd	Pb
AA	1,440 ± 37	2,870 ± 59	27,600 ± 570	1,930 ± 30	101 ± 6	141 ± 5	143 ± 4	101 ± 3	98 ± 4	2.10 ± 0.10	ND
B	1,780 ± 37	2,260 ± 22	27,200 ± 180	1,980 ± 15	51.0 ± 0.5	127 ± 6	54.3 ± 0.7	99.0 ± 2	100 ± 1	2.90 ± 0.03	ND
DD	1,630 ± 76	3,310 ± 94	26,900 ± 170	1,980 ± 23	67.9 ± 2.1	129 ± 6	115 ± 1	104 ± 4	130 ± 2	2.41 ± 0.20	ND
FS	1,770 ± 41	1,260 ± 50	28,000 ± 75	1,460 ± 36	69.6 ± 0.6	129 ± 6	133 ± 7	103 ± 2	161 ± 1	1.66 ± 0.10	ND

AA = Addis Ababa, B = Bichena, DD = Debre Markos, FS = Finote Selam and ND = not detected.

Ten elements (Ca, Mg, Fe, Mn, Cu, Cr, Co, Zn, Ni, and Cd) were detected in both the fennel fruit and soil samples while Pb was not detected in both the fennel fruit and soil samples. The levels of elements however differ significantly among each other and to some extent between sampling sites. The variation between the sampling sites is due to variation in geological, geographical and climatic conditions of the sites. The trend of variation in fennel fruit sample in Addis Ababa site is Ca > Mg > Fe > Cu > Cr > Co > Mn > Zn > Ni > Cd, in Bichena site is Ca > Mg > Fe > Cr > Mn > Cu > Zn > Co > Ni > Cd, in Debre Markos site is Ca > Mg > Fe > Cr > Cu > Co > Mn > Zn > Ni > Cd and in Finote Selam site is Ca > Fe > Mg > Cr > Co > Zn > Mn > Ni > Cu > Cd. The trend of variation in soil sample in Addis Ababa site is Fe > Mg > Mn > Ca > Co > Cr > Cu = Zn > Ni > Cd, in Bichena site is Fe > Mg > Mn > Ca > Cr > Ni > Zn > Co > Cu > Cd, in Debre Markos site is Fe > Mg > Mn > Ca > Ni > Cr > Co > Zn > Cu > Cd and in Finote Selam site is Fe > Ca > Mn > Mg > Ni > Co > Cr > Zn > Cu > Cd. In all the cases concentration of Ca and Fe are much higher than the other elements in the both the fennel fruit and soil, respectively. In both the fennel fruit and the soil the level of Cd is much low at trace level but its concentration in the soil is relatively higher than the fennel fruit.

Table 5 indicates that except Cd which has lowest concentration in samples from Finote Selam and Ni in samples from Addis Ababa all the other essential trace elements lowest concentrations were found in samples from Bichena. All the trace elements have higher concentrations in samples from Addis Ababa (Cu, Cr, Co and Cd) except Zn in samples from Debre Markos and Ni in samples from Finote Selam. These go with the expectation of minor pollution in the capital city, Addis Ababa. Table 6 indicates that Co and Cr was the most accumulated trace element in the soil sample while the level of Cd was the lowest among the elements. Since Ethiopia is 10th in the world and 1st in Africa in cattle, leather industries are flourishing in the country. Relatively larger accumulation of Cr in the soil may be due to release of Cr in the nearby leather industries but the source of Co is unknown.

4. DISCUSSION

Elements are persistent in the environment and tend to bioaccumulate in plants and organisms. Bioaccumulation is of course a normal and essential process enabling the organism to have reserve for latter use for metalloproteins or cofactors or protect themselves against toxic effects (Kitata and Chandravanshi 2012). In this study the bioaccumulation of both trace and major elements were attempted to be verified by roughly comparing their concentration in one of the growth media, soil and those in plants.

In comparison between soil and fennel fruit sample except for Ca fennel fruit showed less or comparable accumulation of elements found in the soil. Higher accumulation was observed for Ca only (Table 7 and Figure 1A). This may be because it is a seasonal plant and the soil is acidic. Ca accumulation ability of fennel fruit makes it valuable Ca resource for infants, pregnant women and elderly persons who need it very much in addition to its medicinal and spicy nature. However, it is not possible to estimate the daily intake of Ca from fennel fruit by infant, pregnant women and elderly persons living in different parts of the country due to their variable daily intake. Also there is no systematic data available on the daily intake of fennel fruit. For Mg and Cu fennel fruit showed comparable accumulation to that of the soil but for Fe, Mn, Cr, Co, Zn, Ni and Cd it showed low accumulation than available in the soil in all the sites. This might indicates that they are not found in bio-available form or fennel fruit has poor accumulation for these elements. The very low concentration of Cd and not detectable Pb in the fennel fruit indicate that the risk of toxicity of these elements is low when consuming it in row or processed forms. The FAO/WHO dietary recommendations of some of the metals are given in Table 8.

The accumulation factor of the fennel fruit is evaluated and the result is presented in Table 7. Ca has highest accumulation coefficient (12.9–14.2) in all the sites. The reason for higher concentration of Ca in the fennel fruit is due to the acidity of the soil as a result of higher concentration of Fe (28,000–26,900 μg/g). The least accumulation is seen for Mn (0.015–0.056) in all the sites. Elements like Mg, Cu and Cd have variable concentration in the plant as that in the soil. For Cr, moderate difference in concentration is observed. In the case of Zn and Co, the concentration in the soil is higher than the fennel fruit at roughly by half. For other elements like Fe, Mn and Ni the concentration of the soil is very high (Figure 1A and B).

Table 7. Accumulation coefficient of elements from the soil to fennel fruit

Element	Sampling sites			
	AA	B	DM	FS
Ca	14.2	12.9	13.1	12.9
Mg	0.72	0.87	1.2	0.96
Mn	0.021	0.015	0.056	0.035
Fe	0.041	0.070	0.055	0.052
Zn	0.35	0.38	0.43	0.40
Cu	1.0	1.4	0.35	0.26
Cr	0.67	0.46	0.76	0.40
Co	0.49	0.19	0.11	0.54
Ni	0.25	0.19	0.19	0.15
Cd	0.91	0.58	0.78	0.96

AA = Addis Ababa, B = Bichena, DD = Debre Markos and FS = Finote Selam.

4.1 Comparison of levels of elemental in fennel fruit with literature values

The major elements (Ca, Mg, Fe and Mn) have been determined in the fennel fruit in different countries from different points of view (nutritional, health problems, crop yield, etc.). Some of the available studies are presented in Table 9 along with the present study for comparison.

The level of Ca, in this study is comparable with the study from India (Kumar et al. 2005) but much higher than from Turkey (Başgel and Erdemoğlu 2006; Ozcan and Akbulut 2007). The level of Mg in this study is lower than from Turkey (Başgel and Erdemoğlu 2006; Ozcan and Akbulut 2007), and much lower than study from India (Kumar et al. 2005). The level of Fe in this study is much higher than studies from India and Turkey (Başgel & Erdemoğlu 2006; Ozcan and Akbulut 2007; Kumar et al. 2005). The level of Mn in this study is lower than study from India but higher than Turkey (Başgel and Erdemoğlu 2006; Ozcan and Akbulut 2007;

Kumar et al. 2005). Generally the values are acceptable except Fe which is very high than the literature values.

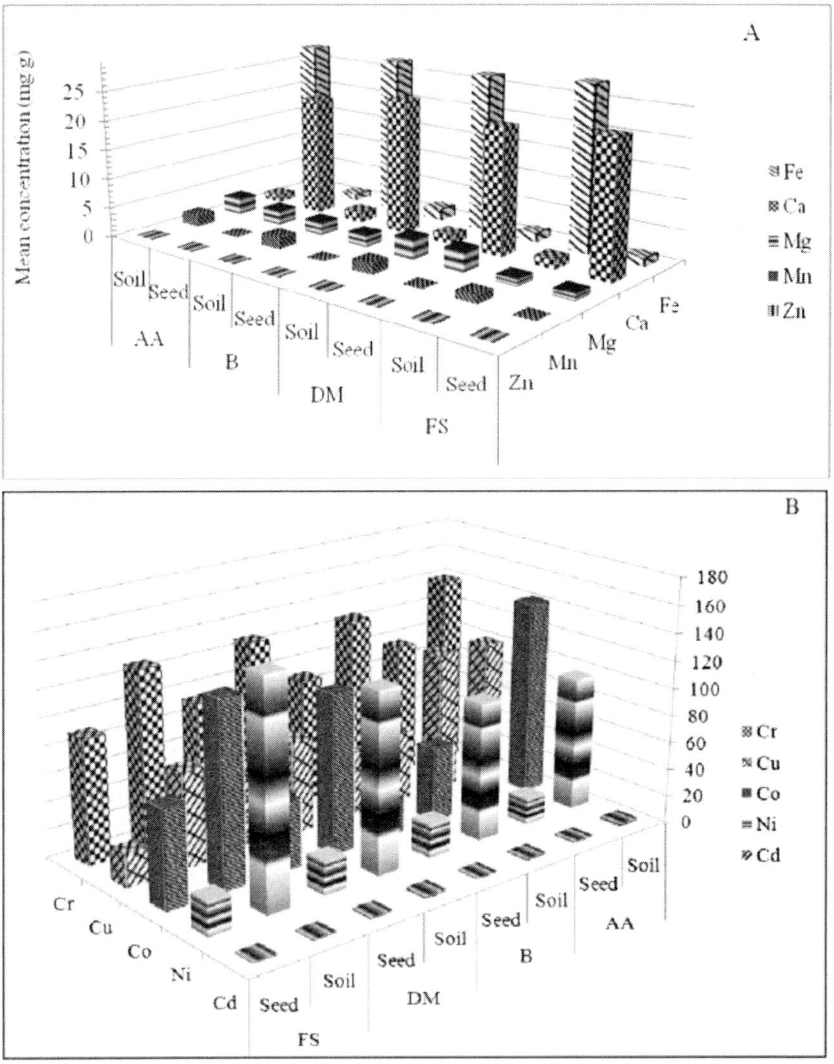

Figure 1. Comparison of concentration of elements in the fennel fruit to elements in the soil. A**. Major elements and** B**. trace elements. AA = Addis Ababa, B = Bichena, DD = Debre Markos and FS = Finote Selam.**

Table 8. FAO/WHO dietary recommendation of some of the minerals (mg/day)

Metals	Population groups			
	Infants	Pregnant women	Lactating women	Elderly (>65 year)
Ca	300-400	1200	1000	1300
Mg	26-36	220	270	190-230
Zn	2.8-4.1	5.5-10	7.2-9.5	4.9-7.0
Fe	8-9	1040	15	11-14

N.B. WHO recommendation for toxic metals in herbal medicines and products are: 10 mg/kg and 0.3 mg/kg for Pb and Cd, respectively.

Comparisons of levels of trace elements in fennel fruit reported in various parts of the world are summarized in Table 10. Results for Cu, Cr, and Co of this study are much higher than the literature values given in Table 10. This variation may be due to difference in the geographical location, climate, agricultural chemicals like fertilizers or errors in sample preparation and analysis. Extremely higher values for Cr may be related to pollution from the tannery industry found around the sampling areas. Results of this study for Zn, Ni and Cd are comparable with literature values.

4.2 Multivariate analysis of variance

Variation in the mean levels of elements between the fennel fruit and soil samples were tested based on Tukey HSD multiple comparison model. Mg and Cd show significant differences ($p < 0.05$) at 95% confidence levels for all sampling sites whereas Zn in all sites and Fe except Addis Ababa and Debre Markos sites showed significant similarity ($p > 0.05$). These may indicate that fennel fruit has special storing capability for Mg and Cd irrespective of their concentration in the soil whereas Fe and Zn accumulation is the reflection of their concentration in the soil. Other elements show partial significant similarity and difference.

Table 9. Summary of elemental levels reported fennel fruit worldwide (all values are in µg/g)

Country	Ca	Mg	Fe	Mn	Ref.
India	23,900 ± 1300	5,110 ± 420	744 ± 20	72.3 ± 3.0	(Kumar et al. 2005)
Turkey	10,800 ± 220	2,770 ± 39	225 ± 9	27.8 ± 0.4	(Başgel & Erdemoğlu 2006)
Turkey	6,750 ± 120	3,400 ± 380	316 ± 35	33.4 ± 5.3	(Ozcan and Akbulut 2007)
Pakistan	-	-	1030 ± 290	877 ± 85	(Khattak and Khattak 2011)
Ethiopia	21,900 ± 1,190	2,280 ± 900	1,500 ± 310	44.0 ± 9.2	Present study

Table 10. Summary of trace elemental levels in fennel fruit worldwide (all values in μg/g)

Country	Cu	Cr	Co	Zn	Ni	Cd	Ref.
India	10.9 ± 0.8	1.50 ± 0.35	0.54 ± 0.10	34.5 ± 4.4	1.46 ± 3.43	0.086 ± 0.034	(Kumar et al. 2005)
Turkey	16.2 ± 0.4	1.04 ± 0.20	0.40 ± 0.02	37.0 ± 2.4	5.40 ± 0.14	0.004 ± 0.001	(Başgel & Erdemoğlu 2006)
Turkey	8.28 ± 0.85	35.9 ± 9.12	-	20.8 ± 4.78	28.7 ± 1.38	0.50 ± 0.1	(Ozcan and Akbulut 2007)
Pakistan	117 ± 22	-	-	37.5 ± 3.0	-	-	(Khattak and Khattak 2011)
Pakistan				24.7 ± 0.5			(Chowdhary et al. 2014)
Ethiopia	59.5 ± 33.1	93.5 ± 3.1	53.5 ± 20.5	40.1 ± 3.5	22.8 ± 2.7	1.54 ± 0.16	Present study

Ca (only in Bichena and Finote Selam), Cu (only in Finote Selam and Debre Markos), Co (only in Addis Ababa and Finote Selam) and Ni (only in Addis Ababa and Bichena) fennel fruit and soil showed significant similarity whereas for the other they showed significant difference. On the other hand, Mn and Cr showed half-half significant similarity and difference.

4.3 Pearson correlation of elements

The high association between elements, evidenced by high positive correlation coefficient, can arise from common anthropogenic or natural sources as well as from similarity in chemical properties. The Pearson correlation matrices using correlation coefficient (r) for the fennel fruit samples of the four sits are shown in Table 11. There is poor negative relation between Ca and Mn, Mg and Fe, Mg and Cu, Mg and Cr, Mg and Ni, Mg and Cd, Mn and Zn, Fe and Zn, Fe and Ni, Zn and Cu, Zn and Cr, Cu and Ni, Cr and Co and Cr and Cd. Poor positive correlation was found between Ca and Fe, Zn, Co, Ni, and Cd; Mn and Cu, Cr, Co and Cd; Fe and Cd; Cu and Co and Cd; and Ni and Cr and Co. Mg has no poor positive relationship with any of the elements. Strong negative relation was seen between Ca and Cr; Mg and Mn and Co; Mn and Ni; Fe and Cr; Zn and Co, Ni and Cd; and Cu and Cr. Strong positive relationship found between Ca and Mg and Cu; Mg and Zn; Mn and Fe; Fe and Cu and Co; and Cd and Co and Ni.

The Pearson correlation matrices using correlation coefficient (r) for the fennel fruit and soil samples of the four sits are shown in Table 12. There is poor relation for Mg, Cu, Cr, and Co in Addis Ababa site, Ca, Fe, Co and Ni in Bichena site and Ni and Cd in Debre Markos site between elements in the fennel fruit and soil. There is no relation between Mg in the soil and fennel fruit in Bichena site which might imply different sources. The rest elements have good relation (at least greater than ± 0.5) between elements in the fennel fruit and soil.

Table 11. Single tailed Pearson Correlation of elements in the fennel fruit (N = 12)

Elements	Ca	Mg	Mn	Fe	Zn	Cu	Cr	Co	Ni	Cd
Ca	1.000	0.553▲	**−0.268**	0.402•	0.232•	0.588▲	−0.964	0.192•	0.118•	0.240•
Mg	0.553	1.000	−0.605	**−0.313**	0.768▲	**−0.047**	**−0.427**	−0.652	**−0.152**	−0.521
Mn	−0.268	−0.605	1.000	0.759▲	**−0.173**	0.442•	0.005•	0.345•	−0.587	0.185•
Fe	0.402	−0.313	0.759	1.000	**−0.154**	0.820▲	−0.617	0.585▲	**−0.352**	0.467•
Zn	0.232	0.768	−0.173	−0.154	1.000	**−0.048**	**−0.224**	−0.825	−0.663	−0.801
Cu	0.588	−0.047	0.442	0.820	−0.048	1.000	**−0.735**	0.410•	**−0.280**	0.312•
Cr	−0.964	−0.427	0.005	−0.617	−0.224	−0.735	1.000	**−0.266**	0.064•	**−0.265**
Co	0.192	−0.652	0.345	0.585	−0.825	0.410	−0.266	1.000	0.490•	0.941▲
Ni	0.118	−0.152	−0.587	−0.352	−0.663	−0.280	0.064	0.490	1.000	0.604▲
Cd	0.240	−0.521	0.185	0.467	−0.801	0.312	−0.265	0.941	0.604	1.0

AA = Addis Ababa, B = Bichena, DD = Debre Markos, FS = Finote Selam.

Bold font indicates poor negative relation (r > −0.500).

The • indicate poor positive relation (r < 0.500).

The ▲ indicate strong positive relation (r > 0.500).

Table 12. Correlation of elements in the fennel fruit and soil samples in the four sits (n = 12)

Site/Elements/	Ca	Mg	Fe	Mn	Cu	Cr	Co	Zn	Ni	Cd
AA	−0.842	0.425•	−0.663	0.985▲	**−0.290**	0.464•	0.300•	−0.993	0.683▲	0.500▲
B	**−0.395**	**0.000***	**−0.092**	0.947▲	0.615▲	0.957▲	**−0.366**	0.704▲	**−0.211**	−0.500
DM	−0.613	−0.707	0.979▲	0.756▲	0.936▲	−0.699	0.977▲	−0.998	0.219•	**−0.013**
FS	−0.721	−0.689	0.569▲	−0.966	−0.693	−0.950	−0.500	0.862▲	−0.984	−0.619

AA = Addis Ababa, B = Bichena, DD = Debre Markos, FS = Finote Selam.

Bold font indicates poor relation (r > – 0.500).

The * indicate no relation.

The • indicate poor positive relation (r < 0.500).

The ▲ indicate strong positive relation (r > 0.500).

5. CONCLUSION

In this study fennel fruit and the soil in which the plant was grown was analyzed for major and trace elements (Ca, Mg, Fe, Mn, Cu, Cr, Co, Zn, Ni, Cd and Pb). The fennel fruit was found to have higher concentration of Ca, moderate concentration of Mg and Fe, and lower but considerable concentration of Cr, Cu, Co, Mn and Zn; relatively low concentration of Ni and very low concentration of Cd. The soil samples contain the highest concentration of Fe followed by Mg, Ca Mn, Co, Cr, Zn, Ni, and Cu. From the trace elements Cu showed higher accumulation factor. The concentration of Cd was higher in the soil than in the fennel fruit. Concentration of Pb in both the soil and the fennel fruit was below the method detection limit. For all the elements higher or comparable concentration were found in the soil and the fennel fruit except Ca in all the sites and Cu in Addis Ababa and Bichena sites.

The Ethiopian fennel fruit contain higher amount of Ca, and considerable amount of other micronutrients. It contains only trace level of Cd and Pb was not detected which showed that it is safe for human consumption in daily life. These results may also indicate that fennel plant has lesser tendency to accumulate toxic elements in the fruit.

ACKNOWLEDGEMENTS

The authors express their gratitude to the Department of Chemistry, Addis Ababa University, Ethiopia, for providing the laboratory facilities. FDE is thankful to the Wolayta Sodo University, Wolayta Sodo, Ethiopia, for sponsoring his study.

Research and Graduate Programs of Addis Ababa University, Ethiopia had no role in the design, data collection, analysis or publication of the manuscript.

AUTHORS' CONTRIBUTIONS

FDE contributed to the sampling, analysis, and design of the study. FDE and BSC contributed to the interpretation of data and drafting the manuscript. BSC contributed to critical revision of the paper. All authors have read the manuscript and gave their final approval for this version of the manuscript to be published.

REFERENCES

1. Başgel S, Erdemoğlu SB: **Determination of mineral and trace elements in some medicinal herbs and their infusions consumed in Turkey.** *Sci Total Environ* 2006, **359:**82–89.

2. Cetin B, Ozer H, Cakir A, Polat T, Dursun A, Mete E, Ozturk E, Ekinci M: **Antimicrobial activities of essential oil and hexane extract of florence fennel [*Foeniculum vulgare* var. azoricum (Mill.) Thell.] against food borne microorganisms.** *J Med Food* 2010, **13:**196–204.

3. Chéry CC, Herremans S, Lierde V, Vanhaecke F, Freitas MC, Jasekera R: **Ultra-micro trace element contents in spices from Sri Lanka.** *J Radioanal Nucl Chem* 2008, **276:**143–147.

4. Chowdhary F, Bukhsh A, Chaman S, Naeem M: **Detection and measurement of zinc concentrations in selected indigenous medicinal plants of Pakistan.** *Adv Pharm Ethnomed* 2014, **2:**10–13.

5. Dadalioglu I, Evrendilek GA: **Chemical compositions and antibacterial effects of essential oils of Turkish oregano (*Origanum minutiflorum*), bay laurel (*Laurus nobilis*), Spanish lavender (*Lavandula stoechas* L.), and fennel (*Foeniculum vulgare*) on common foodborne pathogens.** *J Agric Food Chem*2004, **52:**8255–8260.

6. Demirel S, Tuzen M, Saracoglu S, Soylak M: **Evaluation of various digestion procedures for trace element contents of some food materials.** *J Hazard Mater* 2008, **152:**1020–26.

7. Desideri D, Meli MA, Roselli C, Feduzi L: **Determination of essential and non-essential elements in herbal tea and camomile by polarised X-rays fluorescence spectrometer (EDPXRF).** *J Radioanal Nucl Chem* 2011,**290:**391–396.

8. Diao W-R, Hu Q-P, Zhang H, Xu J-G: **Chemical composition, antibacterial activity and mechanism of action of essential oil from seeds of fennel (*Foeniculum vulgare* Mill.).** *Food Control* 2014, **35:**109–116. 1

9. Divrikli U, Horzum N, Soylak M, Elci L: **Trace heavy metal contents of some spices and herbal plants from western Anatolia, Turkey.** *Int J Food Sci Technol* 2006, **41:**712–16.

10. El-Motaium RA, El-Seoud MAA: **Irradiated sewage sludge for production of fennel plants in sandy soil.***Nutr Cycl Agroecosyst* 2007, **78:**133–142.

11. FAO/WHO: *Human Vitamin and Mineral Requirements, report of a joint FAO/WHO expert consultation*. 2001.

12. Faudale M, Viladomat F, Bastida J, Poli F, Codina C: **Antioxidant activity and phenolic composition of wild, edible, and medicinal fennel from different Mediterranean countries.** *J Agric Food Chem* 2008, **56:**1912–1920.

13. Getahun A: *Some common medicinal and poisonous plants used in Ethiopian folk medicine*. Addis Ababa University press, Addis Ababa, Ethiopia; 1976.

14. He W, Huang B: **A review of chemistry and bioactivities of a medicinal spice: *Foeniculum vulgare* .** *J. Med. Plant Res* 2011, **5:**3595–3600.

15. Javidnia K, Dastgheib L, Mohammadi Samani S, Nasiri A: **Antihirsutism activity of fennel (fruits of*Foeniculum vulgare*) extract – a double-blind placebo controlled study.** *Phytomedicine* 2003, **10:**455–458.

16. Khattak MI, Khattak MI: **Study of heacy trave metals in some medicinal-herbal plants of Pakistan.** *Pakistan J Bot* 2011, **43:**2003–2009.

17. Kim SI, Chang KS, Yang YC, Kim BS, Ahn YJ: **Repellency of aerosol and cream products containing fennel oil to mosquitoes under laboratory and field conditions.** *Pest Manag Sci* 2004, **60:**1125–1130.

18. Kitata RB, Chandravanshi BS: **Concentration levels of major and trace metals in onion (*Allium cepa* L.) and irrigation water around Meki Town and Lake Ziway, Ethiopia.** *Bull Chem Soc Ethiop* 2012, **26:**27–42.

19. Kumar A, Nair AGC, Reddy AVR, Garg AN: **Analysis of essential elements in Pragya-peya a herbal drink and its constituents by neutron activation.** *J Pharm Biomed Anal* 2005, **37:**631–638.

20. Mallinckrodt MG-v, Meissne D: **General aspects of the role of metals in clinical chemistry.** In *Handbook on Metals in Clinical and Analytical Chemistry*. Edited by: Seiler HG, Sigel A, Sige H. New York: Marcel Dekker; 1994:4–27.

21. Mohamad RH, El-Bastawesy AM, Abdel-Monem MG, Noor AM, Al-Mehdar HA, Sharawy SM, El-Merzabani MM:**Antioxidant and anticarcinogenic effects of methanolic extract and volatile oil of fennel seeds (*Foeniculum vulgare*).** *J Med Food* 2011, **14:**986–1001.

22. O'Dell BL, Sunde RA: *Handbook of Nutritionally Essential Mineral Elements Clinical Nutrition in Health and Disease*. CRC Press, New York; 1997.

23. Oktay M, Gülçin İ, Küfrevioğlu Öİ: **Determination of in vitro antioxidant activity of fennel (*Foeniculum vulgare*) seed extracts.** *LWT - Food Sci Technol* 2003, **36:**263–271.

24. Ozcan MM, Akbulut M: **Estimation of minerals, nitrate and nitrite contents of medicinal and aromatic plants used as spices, condiments and herbal tea.** *Food Chem* 2007, **106:**852–858.

25. Ozcan MM, Unver A, Ucar T, Arslan D: **Mineral content of some herbs and herbal teas by infusion and decoction.** *Food Chem* 2008, **106:**1120–1127.

26. Parthasarathy VA, Chempakam B, Zachariah TJ: *Chemistry of Spices*. CAB International, UK; 2008.

27. Qiu J, Li H, Su H, Dong J, Luo M, Wang J, Leng B, Deng Y, Liu J, Deng X: **Chemical composition of fennel essential oil and its impact on *Staphylococcus aureus* exotoxin production.** *World J. Microbiol Biotechnol*2012, **28:**1399–1405.

28. Rahimi R, Ardekani M: **Medicinal properties of *Foeniculum vulgare* Mill. in traditional Iranian medicine and modern phytotherapy.** *Chin J Integr Med* 2013, **19:**73–79.

29. Rasul A, Akhtar N, Khan BA, Mahmood T, Uz Zaman S, Khan HM: **Formulation development of a cream containing fennel extract: in vivo evaluation for anti-aging effects.** *Pharmazie* 2012, **67:**54–58.

30. Rather MA, Dar BA, Sofi SN, Bhat BA, Qurishi MA: ***Foeniculum vulgare* : A comprehensive review of its traditional use, phytochemistry, pharmacology, and safety.** *Arabian J. Chem* 2012.

31. Renna M, Gonnella M: **The use of the sea fennel as a new spice-colorant in culinary preparations.** *Int J Gastronomy Food Sci* 2012, **1:**111–115.

32. Sembratowicz I, Rusinek-prystupa E: **Effects of brewing time on the content of minerals in infusion of medicinal herbs.** *Polish J Environ Stud* 2014, **23:**177–186.

33. Senatore F, Oliviero F, Scandolera E, Taglialatela-Scafati O, Roscigno G, Zaccardelli M, De Falco E: **Chemical composition, antimicrobial and antioxidant activities of anethole-rich oil from leaves of selected varieties of fennel [*Foeniculum vulgare* Mill. ssp. vulgare var. azoricum (Mill.) Thell].** *Fitoterapia* 2013,**90:**214–219.

34. Shahat AA, Ibrahim AY, Hendawy SF, Omer EA, Hammouda FM, Abdel-Rahman FH, Saleh MA: **Chemical composition, antimicrobial and antioxidant activities of essential oils from organically cultivated fennel cultivars.** *Molecules* 2011, **16:**1366–1377. 10.3390/molecules 16021366

35. Shirahatt RV, Ankola AV, Nagesh L: **Effect of fennel seeds on dental plaque and salivary pH – a clinical study.** *J Oral Health Community Dent* 2010, **4:**38–41.

36. Singh V, Garg AN: **Availability of essential trace elements in Indian cereals, vegetables and spices using INAA and the contribution of spices to daily dietary intake.** *Food Chem* 2006, **94:**81–89.

37. Szymczycha-Madeja A, Welna M, Pohl P: **Elemental analysis of teas and their infusions by spectrometric methods.** *Trends Anal Chem* 2012, **35:**165–181.

38. Zengin M, Özcan MM, Çetin Ü, Gezgin S: **Mineral contents of some aromatic plants, their growth soils and infusions.** *J Sci Food Agric* 2008, **88:**581–589.

39. Zuberi MI, Kebede B, Gosaye T, Belachew O: **Species of herbal spices grown in the poor farmers' home gardens of West Shoa, Highlands of Ethiopia: an ethnobotanical account.** *J Biodivers Environ Sci* 2014,**4:**164–185.

Chapter 7

Seed wintering and deterioration characteristics between weedy and cultivated rice

Jung-Sun Baek and Nam-Jin Chung

ABSTRACT

Background

Incidences of weedy rice continuously occurred in paddy fields because its shattering seeds were able to over-winter. In this research, the seed deterioration of weedy rice was investigated compared with cultivated rice, and the wintering characteristics of these two types of rice were investigated with the field wintering test, freezing resistance test, and accelerated aging test.

Results

For the wintering test, the seeds of weedy rice were placed on the soil surface of a paddy with cultivated rice seeds during the 2008/2009 and 2009/2010 winter seasons from November to April. The viability of seeds after wintering was 4.3% for cultivated rice, but 92.7% for weedy rice in

2008/2009. In the second wintering test, the seeds were placed under flooded and dry paddy conditions. The seed viability of cultivated rice was 5% in dry paddy and 0.5% in flooded paddy, but weedy rice maintained a high viability during winter of 90% in the dry paddy and 61% in the flooded paddy. Following freezing treatment of the imbibed seeds, the seed viability was 78% for weedy rice and 16% for cultivated rice. The deterioration of seed tissue induced by freezing treatment was observed by the tetrazolium test. In an accelerated aging test at low temperature and soaking conditions, the seed viability of the weedy rice was 40% higher than the cultivated rice 90 days after treatment. During accelerated aging of seeds, the protein content remained higher in the weedy rice compared to the cultivated rice, and fat acidity remained lower in the weedy rice compared to the cultivated rice. Catalase and superoxide dismutase activity of the weedy rice was 4 times higher than that of the cultivated rice, and DPPH radical scavenging activity of the weedy rice was also much higher than for the cultivated rice.

Conclusion

In conclusion, the superior ability of seed wintering in weedy rice was based on freezing resistibility of embryo cellular tissue and higher antioxidant activity to protect seed deterioration during the winter season.

KEYWORDS

Weedy rice Seed, Wintering, Deterioration, Freezing resistance, Antioxidant, Accelerated aging

1. BACKGROUND

Traditional rice cultivation has been via transplanting of rice seedlings. However, due to the severe problems of agricultural water scarcity and farm labor shortage, direct-seeding methods have been practiced in many parts of the world (Cao et al. [2007]; Savary et al. [2005]; Tabbal et al. [2002]; Tomita et al. [2003]). In most countries, weedy rice infestation increased significantly after shifting from rice transplanting to direct seeding, and it was recognized as a noxious weed (Ottis et al. [2005]; Suh [2003]; Vaughan et al. [2001]). Weedy rice infestations have been reported to have spread to 40–75% of the total area of rice cultivation in Europe, 40% in Brazil, 55% in Senegal, 80% in Cuba, and 60% in Costa Rica (Fogliatto et al. [2010]).

The occurrence of weedy rice has very large effects on rice yield (Pantone and Baker [1991]; Pantone et al. [1992]). Weedy rice infestations are responsible for significant yield losses, which are particularly severe in short varieties and late planting cultivation (Fogliatto et al. [2010]). Weedy rice is so productive that it can spread and cause major economic damage (Kane and Baack [2007]), but it is impossible to control weedy rice during the rice cultivation period by herbicide because it belongs to the same species as the cultivated rice (Delouche et al. [2007]; Suh [2003]).

Weedy rice looks similar to cultivated rice in outward appearance, but has a high level of adapting ability to environmental stress in physiological traits. Being distinguishable by a red pericarp, it disperses immediately by shattering at maturity, guaranteeing its continuation (Chung [2010]; Chung and Ahn [2005]; Chung and Paek [2003]; Ottis et al. [2005]; Suh [2008]). The shattering nature of weedy rice after ripening makes it difficult to be removed when cultivated rice is harvested. The occurrence of weedy rice becomes more severe the following year due to shattered seed present in the soil (Seong et al. [2004]).

The reason of continuous incidences of weedy rice was that its shattering seeds were able to over-winter. In this respect, the deterioration of weedy rice seeds during winter might be different with cultivated rice in freezing resistance and antioxidant activities. There were no reports on the freezing resistance, but some reports of antioxidant activities related with seed deterioration in rice seeds (Bailly [2004]). Generally, it has become increasingly accepted that damage resulting from reactive oxygen species (ROS) or oxidative stress plays a role in the seed aging process. ROS are highly reactive and may modify and inactivate proteins, lipids, DNA, and RNA and induce cellular dysfunctions. Plants require high contents of antioxidants such as ascorbate, α-tocopherol, glutathione, and phenolic compounds that can remove ROS as a defense mechanism (Schoner and Heinrich Krause [1990]), and enhance the activities of antioxidant enzymes like superoxide dismutase (SOD), peroxidase (POD), catalase (CAT), and ascorbate peroxidase (APX) (Bowler et al. [1992]). If balances are broken between ROS generation and removal in plant cells, oxidant losses occur along with damage to lipid peroxidation and cell membranes (Larson [1995]).

In the present study, the wintering of shattered seeds of weedy rice on paddy fields was compared with cultivated rice, and seed deterioration and related characteristics concerning seed wintering are reported.

2. RESULTS

Seed viability of weedy rice (WD-3) and cultivated rice (Hopum) after a wintering test on the surface of a paddy field is shown in Figure 1. The viability of wintered seeds was 92.7% for the weedy rice and just 4.3% for the cultivated rice. Wintering characteristics of weedy rice in the winter period may vary depending on the moisture state of the paddy surface. Therefore, the second field experiment for seed wintering was carried out in a flooded paddy as well as in a dry paddy field. The results were almost identical to the first year experiment except that the seed viabilities significantly decreased in the flooded paddy as compared with the dry paddy field for both weedy rice (WD-3, PBR) and cultivated rice (Hopum, Ilpum). As shown in Figure 2, the seed viability of cultivated rice rapidly decreased from December to February regardless of paddy conditions and the final average viabilities after wintering were 5.0% (4.3% in Hopum, 5.7% in Ilpum) in the dry paddy and 0.5% (0.3% in Hopum, 0.7% in Ilpum) in the flooded paddy. For weedy rice, the viability was maintained over 90% in the dry paddy, but decreased slightly during winter to a mean of 61% (WD-3: 76%, PBR: 58%) in the flooded paddy field. At the early stage of wintering, the germination percentage of weedy rice was much lower overall than of cultivated rice. This may be because weedy rice had primary seed dormancy (Bewley and Black [1982]; Cohn and Hughes [1981]), and generally exhibited stronger dormancy than cultivated rice (Gu et al. [2003]). In considering the relationship of seed dormancy to wintering, a second area of interest is whether dormancy increases the lifespan of seeds. In this result, the dormancy of all seeds disappeared in December regardless of its intensity and did not affect the resistant to wintering and freezing stress. Roberts ([1972]) reviewed on the subject and concluded that available evidence was not sufficient to establish even a causal relationship. However, this does not exclude the possibility that the lifespan of dormant seeds of native or wild plants buried in the soil under natural conditions may be longer than the lifespan of non-dormant seeds (Justice and Bass [1978]).

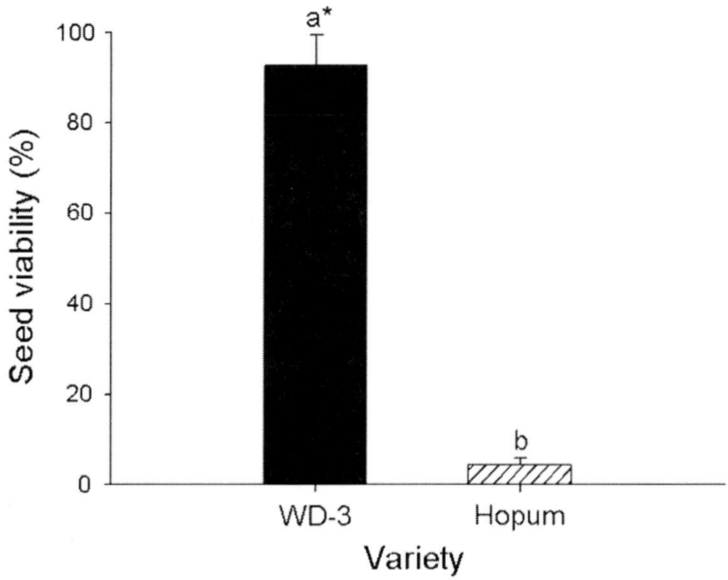

Figure 1. Seed viability of weedy rice (WD-3) and cultivated rice (Hopum) after wintering on the surface of a paddy field. After harvesting Hopum, natural shattering weedy rice (WD-3) and artificially shattering cultivated rice (Hopun) were placed on the soil surface at 3 spots of the paddy field under wire net protection during winter from November 2008 to April 2009. Germination was tested at 25°C for 14 days. *Different letters on the bars indicate significant difference of means at $\alpha = 0.01$ by *t*-test.

Viability of seed wintered on the paddy surface may be affected by temperature changes. In particular, subzero temperatures after rain and snow may freeze the seeds and cause them to die. Therefore, in order to winter safely under subzero temperature, seeds should have resistance to freezing injury for not losing its viability even under soaking conditions. Figure 3 shows seed viability after a number of freezing treatments (1–3 cycles). Seed viability values on average of three treatments of freezing were about 78% (PBR: 72%, WD-3: 85%) in weedy rice and 16% (Hopum: 3%, Ilpum: 29%) in cultivated rice. The results indicated that the weedy rice had much higher resistance than the cultivated rice varieties, and there were also considerable variations among accessions regardless of the number of freezing treatments.

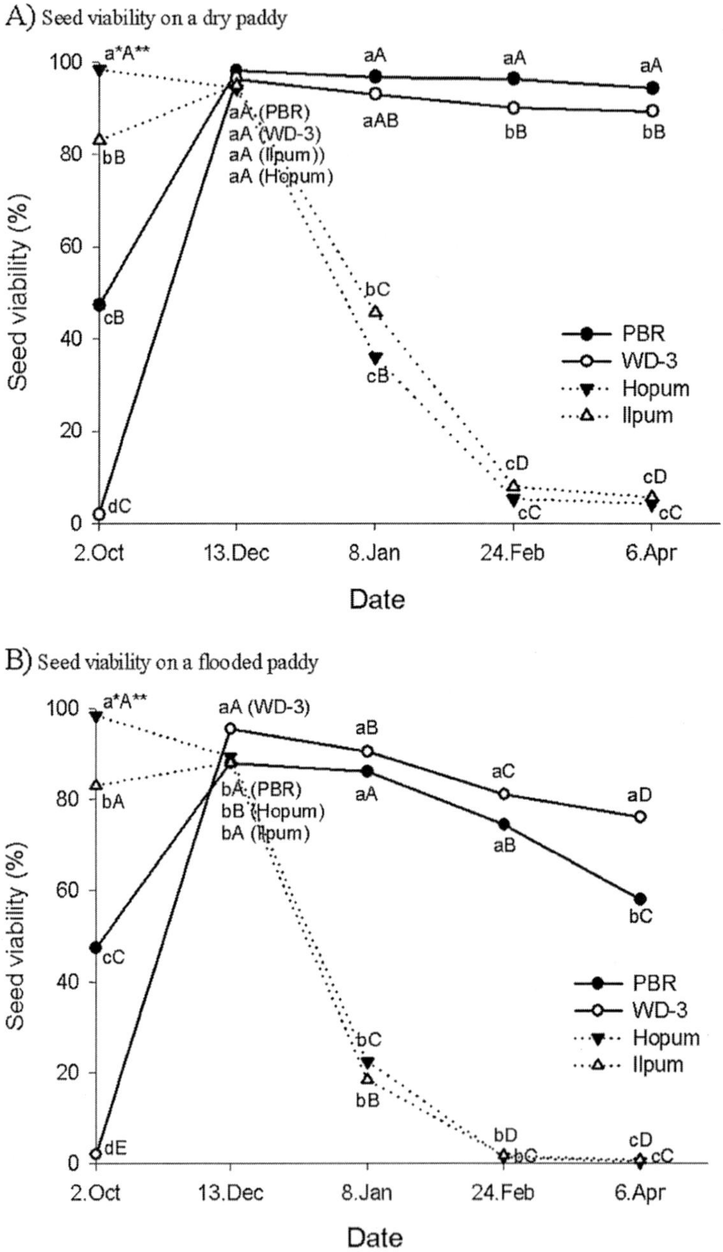

Figure 2. Time course of seed viability during wintering on dry (A) and flooded (B) paddy conditions in weedy rice (PBR, WD-3) and cultivated rice (Hopum, Ilpum). The seeds of 4 varieties (200 g each) were kept on the paddy from November 2009 to April 2010. One was a dry paddy(A) in the same way as the first wintering test and the other was a flooded paddy(B)

in a plastic box (41.0 × 24.5 × 15.0 cm) with 10 cm depth of paddy soil (silty clay loam). During wintering, the seeds were sampled 5 times for a viability test. *Different small letters in a same date (for vertical comparison) indicate significant difference of means at $\alpha = 0.05$ by Duncan's Multiple Range Test. **Different capital letters in the same variety (for horizontal comparison) indicate significant difference of means at $\alpha = 0.05$ by DMRT.

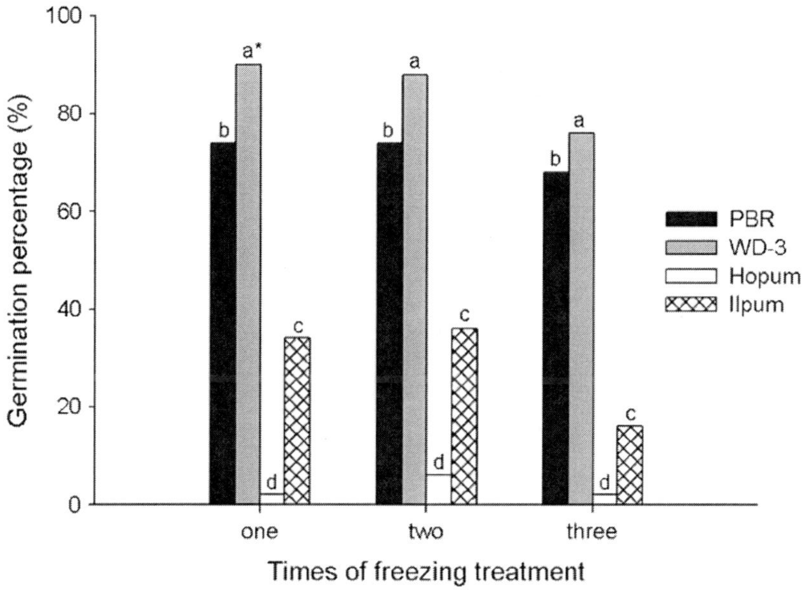

Figure 3. Germination percentage of seeds after a number of freezing treatments in weedy rice (PBR, WD-3) and cultivated rice (Hopum, Ilpum). The seeds were imbibed at 4°C for 24 hours and drained on paper towel. Next, they were frozen at −10°C for 24 hours, and then thawed at 4°C for 24 hours. This was one cycle of freezing treatment. After 1 to 3 cycles of treatment, germination was tested at 25°C for 14 days. . *Different letters on the bars in a treatment indicate significant difference of means at $\alpha = 0.01$ by Duncan's Multiple Range Test.

The deterioration of seed tissue induced by freezing treatment was observed by the tetrazolium test as shown in Figure 4. The frequency of dark purple stained seeds matched up with the freezing resistance of 4 accessions, as shown in Figure 3. In most seeds of weedy rice (PBR and WD-

3), whole embryo tissue and the aleuronic layer were stained dark purple. In the cultivated rice, however, the seeds of Hopum were not completely stained, but those of Ilpum were partially stained, which results indicated that freezing resistance of Ilpum was higher than that of Hopum, and there were variation of freezing resistance among individual seeds in a variety and among rice varieties. Some seeds partially damaged were stained in a light pink at overall tissue such as radicle, plumule, scutellum in embryo and aleurone layer in endosperm. In the deterioration process of seeds under subzero temperature, there was no special tissue structure initiating deterioration.

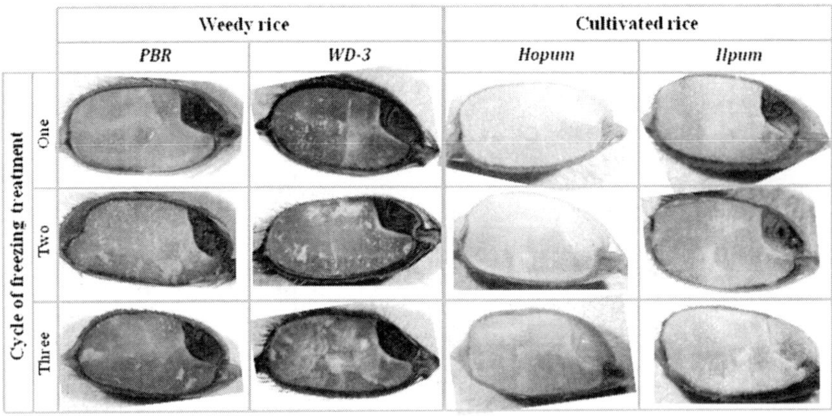

Figure 4. Tetrazolium test of seeds after a number of freezing treatments in weedy rice (PBR, WD-3) and cultivated rice (Hopum, Ilpum). The method of freezing treatment was the same as for Figure 3.

During the winter season, shattered seeds on the paddy surface tended to remain wet at low temperature. Therefore, an accelerated aging test was set up by soaking the seeds at a low temperature (4°C). In this condition, the change of seed viability, physiochemical properties including protein content, fat acidity, and some antioxidant enzyme activities were observed at 10 day intervals during treatment. As shown in Figure 5, the seed viability of weedy rice under the accelerated aging condition started to fall off 50 days after treatment and reached approximately 40% 90 days after treatment. In the cultivated rice, it started to fall off 20 days after treatment, and reached 4% 90 days after treatment.

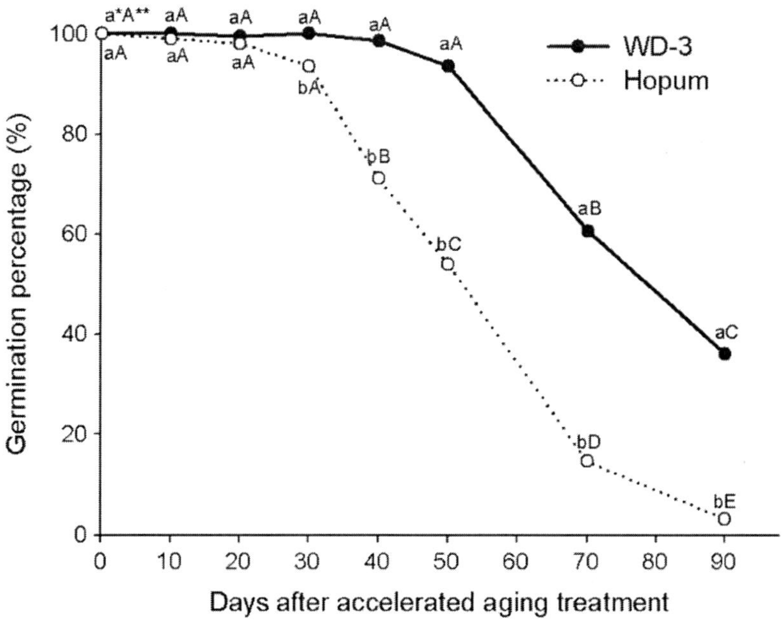

Figure 5. Germination percentage of rice seeds during accelerated aging treatment. For the accelerated aging treatment, the seeds were kept soaking in distilled water at 4°C for 90 days. The seeds were sampled and tested at 10 day intervals. Germination tests were carried out with 4 replicates of 100 seeds at 25°C for 14 days. *Different small letters in a same date (for vertical comparison) indicate significant difference of means at $\alpha = 0.05$ by *t*-test. **Different capital letters in a same variety (for horizontal comparison) indicate significant difference of means at $\alpha = 0.05$ by Duncan's Multiple Range Test.

The time courses of protein content and fat acidity in seeds during accelerated aging treatment are shown in Figure 6. The protein contents of WD-3 and Hopum, which were 7.41% and 6.44% before treatment, respectively, significantly decreased to 5.25% and 4.82% 40 days after treatment, and then maintained at this amount until 90 days after treatment without significant change.

In the case of fat acidity, the initial values were 6.50 KOH mg/100 g in the weedy rice and 6.59 KOH mg/100 g in the cultivated rice, which were the same between the two varieties. Until 40 days after treatment, there were no significant changes of fat acidity, but there was a significant increase and difference between weedy rice and cultivated rice 50 days after treatment. Fat acidities of weedy rice and cultivated rice were 42.9 KOH mg/100 g and 23.9 KOH mg/100 g at day 70, and 44.1 KOH mg/100 g and 30.0 KOH mg/100 g at day 90, respectively.

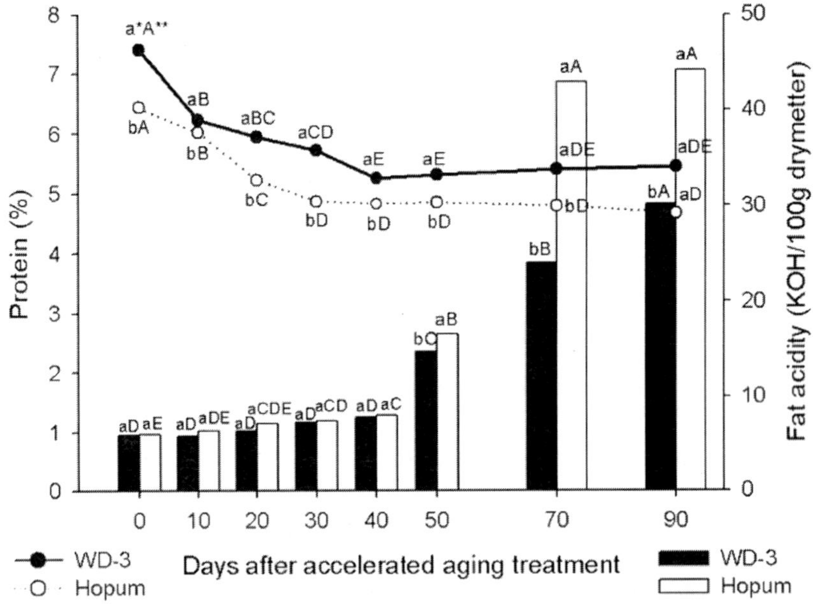

Figure 6. Time course of protein content and fat acidity during accelerated aging treatment for 90 days. The accelerated aging treatment was the same as for Figure 5. * Different small letters in a same date (for vertical comparison) indicate significant difference of means at α = 0.05 by *t*-test. ** Different capital letters in a same variety (for horizontal comparison) indicate significant difference of means at α = 0.05 by Duncan's Multiple Range Test.

The change of superoxide dismutase (SOD) activity during accelerated aging treatment is shown in Figure 7. The initial values of SOD in the seeds were 42.9 units/mg protein in WD-3 and 11.7 units/mg protein in Hopum. Following accelerated aging treatment, SOD activities of seeds increased for 10 days to 104.9 units/mg protein in WD-3 and to 19.9 units/mg protein in Hopum. After peak SOD activity, they decreased in both weedy and cultivated rice until the end of treatment. The change of SOD activity of weedy rice seed rapidly increased and decreased compared to Hopum, but it was maintained at higher levels than that of Hopum. At day 90 of accelerated aging treatment, the SOD activities were 14.3 units/mg protein in weedy rice and 6.4 units/mg protein in cultivated rice. The results of SOD activity changes during accelerated aging suggested that the response of weedy rice seed to environmental stress was more sensitive than cultivated rice.

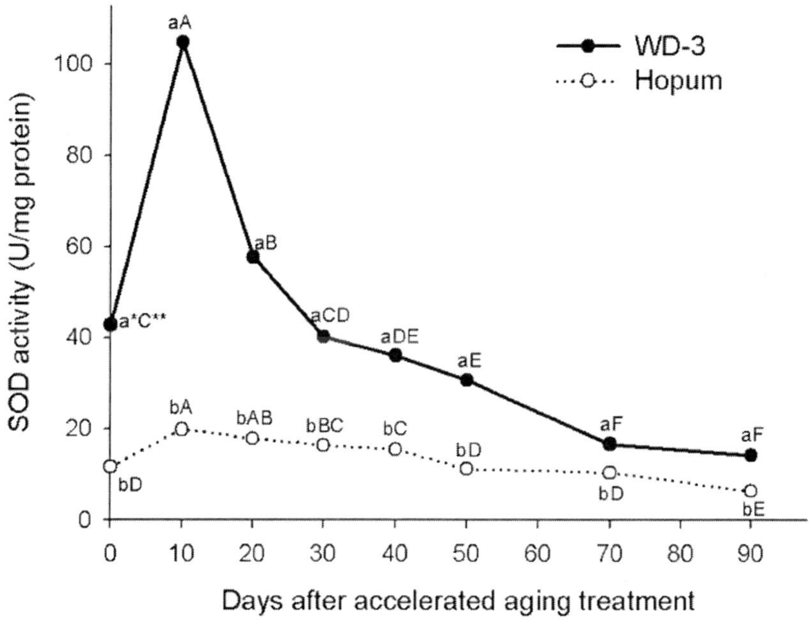

Figure 7. Time course of superoxide dismutase activity during accelerated aging treatment for 90 days. The accelerated aging treatment was the same as for Figure 5. * Different small letters in a same date (for vertical comparison) indicate significant difference of means at α = 0.05 by *t*-test. ** Different capital letters in a same variety (for horizontal comparison) indicate significant difference of means at α = 0.05 by Duncan's Multiple Range Test.

Change in catalase (CAT) activity in the seeds is shown in Figure 8. The initial CAT activity of WD-3 was much higher than that of Hopum. They were 26.9 units/mg protein in weedy rice and 6.6 units/mg protein in cultivated rice. After accelerated aging treatment, CAT activities gradually decreased in both weedy and cultivated rice to 3.9 units/mg protein in weedy rice and 0.8 units/mg protein in cultivated rice at day 90 of accelerated aging treatment. The CAT activity of weedy rice seed persisted at a significantly higher level than that of cultivated rice seed during accelerated aging treatment.

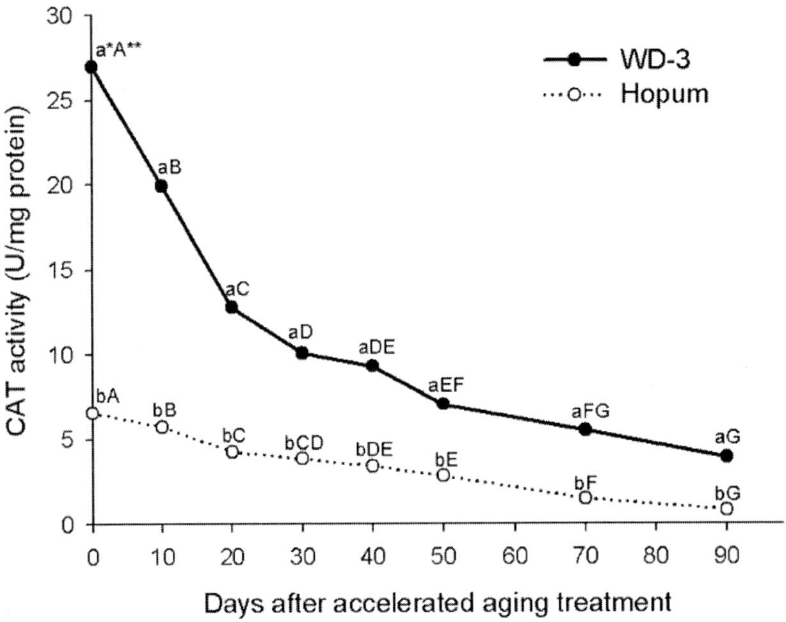

Figure 8. Time course of catalase activity during accelerated aging treatment for 90 days. The accelerated aging treatment was the same as for Figure 5. * Different small letters in a same date (for vertical comparison) indicate significant difference of means at $\alpha = 0.05$ by *t*-test. ** Different capital letters in a same variety (for horizontal comparison) indicate significant difference of means at $\alpha = 0.05$ by Duncan's Multiple Range Test.

Free radical scavenging of phytochemical substances in seeds is an important and generally accepted mechanism for antioxidants inhibiting oxidation (Bandoniene and Murkovic [2002]). The results of seed antioxidant activity measurement during accelerated aging using DPPH is shown in Figure 9. The initial radical elimination activity of WD-3 seed was 93.0%, which was much higher than the 56.6% of Hopum. This persisted during accelerated aging for 90 days without major changes, with approximately 90% in weedy rice and 50% in cultivated rice.

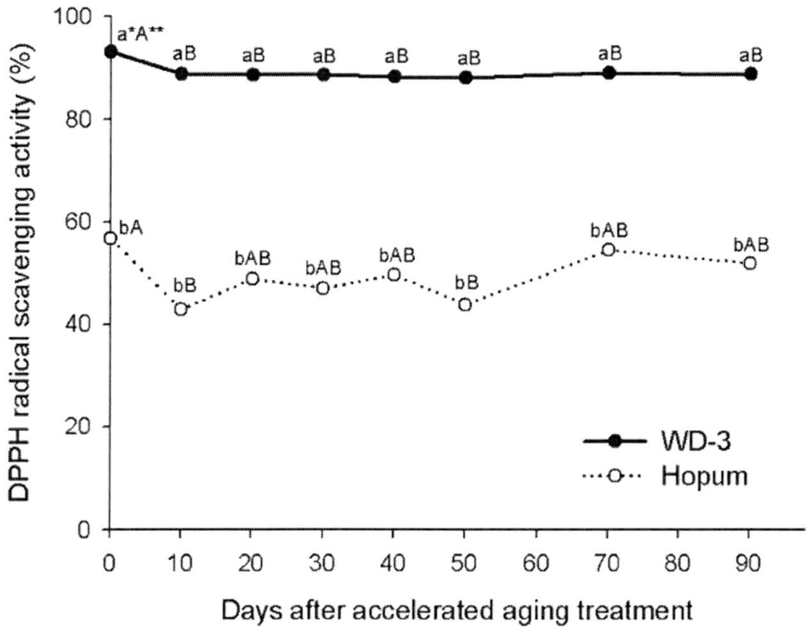

Figure 9. Time course of DPPH radical scavenging activity during accelerated aging treatment for 90 days. The accelerated aging treatment was the same as for Figure 5. * Different small letters in a same date (for vertical comparison) indicate significant difference of means at α = 0.05 by *t*-test. ** Different capital letters in a same variety (for horizontal comparison) indicate significant difference of means at α = 0.05 by Duncan's Multiple Range Test.

3. DISCUSSION

In this study, the wintering rate of weedy rice seeds was about 90% on the dry paddy and 60–80% on the flooded paddy, which was much higher than that of cultivated rice (below 5%). This result was in agreement with several reports about the high germination percentage of weedy rice seeds placed in shallow depths or on the soil surface of paddy fields in the winter (Chung et al. [2000]; Fogliatto et al. [2010]; Ko [2001]; Noldin et al. [2006]; Seong et al. [2004]). Therefore, if a seed of weedy rice spread, it could occur every year and exponentially increase in the rice field. Fogliatto et al. ([2011]) highlighted the significant efficacy of winter flooding in reducing weedy rice infestations in paddy fields. This report is meaningful because the practice showed a significant reduction of weedy rice seed density compared to overwintering under dry conditions, and also winter flooding could favor the decay of seeds of several weeds including weedy rice (Nelms and Twedt [1996]). However, results reported here indicated more than 60% of weedy rice seeds could overwinter under flood conditions. Therefore, winter flooding may not be completely effective but may be an accompanying method to control weedy rice in Korea.

In the winter season, the major factors affecting seed longevity in the field may include precipitation and subzero temperature, as shown in meteorological data in Table 1. When the plant is being frozen gradually, moisture leaves the cell to the outside of the membrane and forms ice crystals causing mechanical damage (Kim [2010]). In the case of seed, if its moisture content is high, it is easily deteriorated in above freezing temperature owing to an increase in respiration and metabolism, and may get damaged in the winter or die at subzero temperature (Baek et al. [2000]).

As shown in Figure 3, the freezing resistance of weedy rice seed was much higher than that of cultivated rice. Weedy rice surviving semi-wildly in rice fields is known to show a high level of adaptability to environmental stress (Chung [2010]; Chung and Ahn [2005]; Suh [2003]). The weedy seeds that shatter on the surface of paddy fields in the autumn should have resistance to freezing in the winter to regenerate in the following germinating season, although there was no report on the freezing resistance of weedy rice seeds. In plants with freezing resistance, antifreeze proteins block the growth of ice crystals in the outside spaces of cells in the tissue, thus preventing cell damage derived from the repetition of freezing and thawing, and also have functions in transforming the generation of ice crystals by adhering to surfaces of ice (DeVries [1986]; Jeong [2009]; Yoo and Hwang [2000]). In this way, antifreeze protein accumulates in the tissue in case of cold acclimation, and thus makes the plant not freeze to death even in freezing temperature. The tetrazolium test in Figure 4 showed that freezing damage of the cultivated rice began in overall tissue of the embryo and aleuronic layer of seed. But, in the weedy rice seeds after freezing treatment, those seed tissues were stained by tetrazolium solution to dark purple to show freezing resistance. Therefore, one hypothesis for a freeze resistance mechanism of weedy rice is that anti-freeze proteins accumulate in the embryo and aleuronic layer of the seed.

Table 1. Meteorological data on temperature and precipitation in Jeonju city during the seed wintering test in the 2008/2009 and 2009/2010 seasons

Year	Month	Temperature (°C)			Days of subzero temperature	Precipitation (mm)	Precipitation days
		Mean	Maximum	Minimum			
2008	Nov.	8.5	14.0	3.8	6	1.4	9
	Dec.	3.1	7.9	−1.7	20	1.1	5
2009	Jan.	−0.4	5.9	−5.1	29	0.9	11
	Feb.	4.2	9.5	−0.6	15	6.0	6
	Mar.	7.0	13.4	1.5	13	4.6	10
	Apr.	12.6	20.2	6.3	1	3.3	5
2009	Nov.	8.6	13.5	4.3	7	2.1	9
	Dec.	1.3	5.7	−2.9	24	2.7	11
2010	Jan.	−0.8	4.1	−5.3	30	3.0	8
	Feb.	3.2	7.7	−0.8	16	7.4	10
	Mar.	6.3	10.9	1.6	9	4.4	14
	Apr.	10.3	16.2	4.8	5	7.0	11

The accelerated aging test has been recognized as an accurate indicator of seed vigor and longevity, and may be generally induced by high temperature and high humidity (Hsua et al. [2003]). The deterioration of seeds wintering on the field surface in temperate regions could be affected by high moisture content at low temperature as discussed earlier, thus accelerated aging can be induced by soaking at low temperature (4°C). The seeds deteriorated rapidly under this accelerated aging method, and there was a large difference in deterioration rate between weedy and cultivated rice. The viabilities of seeds after 90 days of accelerated aging were 4% in cultivated rice and 40% in weedy rice.

The deleterious effect of accelerated aging on the longevity of seeds was associated with the damage occurring to lipid, nucleic acid, and protein owing to oxidative stresses (Cakmak and Horst [1991]; Fujikura and Karssen [1995]; Hsua et al. [2003]; Ito et al. [1999]; McDonald [1992]; Pilar [1998]; Shewfelt and Purvis [1995]; Vartapetian and Jackson [1997]; Yan et al. [1996]). In the accelerated aging test, the difference of protein contents and fat acidities in seeds between weedy and cultivated rice indicated that weedy rice had a higher protective system to aging stress than the cultivated rice.

Scandalios ([1993]) reported that SOD and CAT were effective antioxidant enzymes for decreasing oxidant damage of cells by oxidative stress. The activity of SOD before and after the accelerated aging treatment for 90 days in weedy rice was much higher than in cultivated rice. However, the SOD activities of seeds rapidly increased and decreased with the peak occurring 10 days after treatment. SOD activity decreased if stresses persisted for a long time or excessive stresses were added (Kim and Lee [1992]). CAT is considered a primary enzymatic defense against oxidative stress induced by senescence, chilling, dehydration, osmotic stress, wounding, paraquat, ozone and heavy metals (Kibinza et al. [2011]). Shon et al. ([2008]) reported that CAT activity had exhibited a decreasing tendency upon submerging stress in young rice seedlings, and appeared as the most sensitive enzyme to stress among antioxidant components, regardless of whether it was in seeds. Similarly, CAT activity was reduced after treatment in both the weedy and cultivated rice, where the activity of weedy rice persisted at a higher rate than the cultivated rice. Tanida ([1996]) reported that CAT activity of the embryo in the seed soaked at low temperature had positive correlation with germination percentage after treatment. High SOD and CAT activities, as well as radial scavenging activity were considered to increase the vitality of seeds from excessive stress damage in soaking at low temperature.

4. CONCLUSION

In conclusion, the superior ability of seed wintering in weedy rice was based on the freezing resistance of embryo tissue and the ability of antioxidant enzymes and photochemicals to reduce seed deterioration.

5. METHODS

5.1 Plant materials

Two japonica weedy rice, PBR and WD-3, and two Korean bred cultivars (japonica), Hopum and Ilpum, were used in this experiment. Prior to the laboratory-based experiment, they were regenerated in the experimental field in Chonbuk National University to obtain fresh seeds. The weedy rice were harvested at 30 days after heading, and the cultivated rice were harvested at 50 days after heading. Seeds were air dried until seed moisture content (SMC) became less than 14%. The SMC was measured by oven-drying method. The seeds were sealed in plastic bags, and stored at 4°C until use for about 7–8 months. When they were used, the SMCs were about 12% regardless of varieties.

5.2 Seed wintering test on the paddy field

In the first wintering experiment, Hopum, a bred cultivar, was cultivated in the dry paddy field where the weedy rice (WD-3) occurred at a rate of more than 500 plants per m^2. After harvesting Hopum, naturally shattered weedy rice (WD-3) seeds and artificially shattered cultivated rice (Hopum) seeds were placed on the soil surface at 3 spots of the paddy field under wire net protection during winter from November 2008 to April 2009. Germination tests of seeds collected at 3 spots after wintering were conducted at 25°C for 14 days using 4 replicates of 100 seeds each in accordance with the International Rules for Seed Testing for *Oryza sativa* (ISTA [2009]).

In the second wintering experiment, two weedy rice varieties (WD-3 and PBR) and two Korean cultivars (Hopum and Ilpum) were tested. The seeds of 4 varieties (200 g each) were kept on the paddy from November 2009 to April 2010. The paddy conditions were maintained as two types during winter. One was a dry paddy as was used for the first wintering test and the other was a flooded paddy simulated by plastic boxes (41.0 × 24.5 × 15.0 cm) filled with 10 cm depth of paddy soil (silty clay loam) and superabundant water. During wintering, the seeds (400 grains each from three replicates) were sampled 5 times for viability tests (ISTA [2009]). The meteorological data on temperature, subzero temperature days, precipitation, and precipitation days during the wintering test for 2008/2009 and 2009/2010 is shown in Table 1.

5.3 Seed freezing test

To investigate the resistance of seeds under subzero temperature, the weedy (PBR, WD-3) and cultivated rice (Hopum, Ilpum) were assigned to freezing treatment as follows; the seeds were frozen at −10°C for 24 hours after imbibing at 4°C for 24 hours and then thawed at 4°C for 24 hours. The freezing method was repeated 1 to 3 times, and then the viability of the seeds was tested as stated above in seed wintering test. After freezing treatment, a tetrazolium test was also carried out to investigate the state of seed tissues. Soon after freezing treatment, seeds were kept at 25°C for 15–18 hours. Following that, seeds and embryos were cut vertically using a razor blade, placed in 1% tetrazolium solution (pH 6.5–7.5) at 30°C for 2 hours, and the staining patterns were observed under a dissecting microscope.

5.4 Seed viability and physiochemical test under accelerated aging treatment

To determine the deterioration of soaked seed at low temperature, the seeds of weedy rice (WD-3) and cultivated rice (Hopum) were kept soaking in distilled water at 4°C for 90 days as an accelerated aging treatment, and viability and physicochemical changes of seeds were observed at 10 day intervals. During accelerated aging treatment with three replicates, a 100 g of seeds per replicate was sampled at each sampling time. Among the sampled seeds, 400 grains per replicate was used for the viability test and the rest was ground with liquid nitrogen to powder by auto-mill (TK-AM5, Tokken Inc.) and then dried by freeze dryer (Clean Vac 8, BioTron). The powder was used to analyze protein content, fat acidity, superoxide dismutase (SOD) activity, catalase (CAT) activity and 2,2-diphenyl-2-picrylhydrazyl (DPPH) radical scavenging activity. A germination test was conducted as for the wintering test with four replicates.

Protein content was measured in three replicates by the micro-Kjeldahl method (AOAC [1990]). Seed powder (0.2 g per replicate) was weighed into the Kjeldahl flask. Sulfuric acid and a catalyst (Kjeltabs auto: 1.5 g K_2SO_4, 7.5 mg Se) were added to the flask and the mixture was heated to 420 °C for 1 hour. Total nitrogen of each protein fraction was determined by the automatic Kjeldahl nitrogen analyzer (Kjeltec 2400 Analyzer Unit, Foss Tecator) and protein content was calculated using a conversion factor of 5.95, as shown in the equation below:

$$\text{Protein}(\%) = \text{N}(\%) \times 5.95 \tag{1}$$

Fat acidity (KOH mg/100 g) was analyzed in three replicates by the AOAC ([1970]) method. It was determined by the amount of KOH required to reduce free fat acid included in 100 g of dry material. A total of 10 g of seed powder was placed in a 100 ml flask followed by addition of 50 ml of benzene and 30 min. agitation (150 rpm, 25°C). The mixed solution was filtered through filter paper (Adventec No. 2, 90 mm), and 15 ml of the solution was put in a 100 ml flask and treated with 15 ml phenolphthalein melted in ethyl alcohol. The solution was finally titrated with 0.0045 N-KOH. Ending point of the reaction was determined when color of the solution turned pink, which was the standard color. Tests of the blank and standard varieties were carried out at the same time. Fat acidity was calculated by the following formula (AOAC [1970]):

$$\text{Fat acidity value} = (\text{T} - \text{B}) \times 8.33 \times 100/(100 - \text{W}) \tag{2}$$

T : amount of KOH (ml) used in sample

B : amount of KOH (ml) used in blank

W : moisture content of sample

8.33 : conversion factor according to sample amount and KOH concentration

For analysis of SOD and CAT activity, 1 g of seed powder was homogenized with 5 ml 0.1 M potassium phosphate buffer (pH 7.0) contains 1 mM EDTA and 1 mM DMSO. The homogenized samples were centrifuged at 30,000 × *g* for 20 min. at 4°C. The supernatant was stored at −80°C and used as a crude enzyme extract (Shon et al. [2008]). For SOD activity test by the nitro blue tetrazolium (NBT) reduction method (Beyer [1987]; Shon et al. [2008]), 0.1 ml of 400 mM methionine, 0.1 ml of 2.5 mM NBT and 0.3 ml of 1 mM EDTA were vortexed in the glass test tube at 25°C for 3 min. for temperature balance. Then, 0, 20, 40, and 60 µl of enzyme extracts and 2.45, 2.43, 2.41 and 2.39 ml of 50 mM potassium phosphate buffer (pH 7.8) were added respectively to make final volume of 3 ml of enzyme and potassium phosphate buffer. 50 µl of 120 µM of riboflavin was added to each test tube. The test tubes were softly shaken and placed under fluorescent lamps for 15 min., and absorbance was measured at 560 nm by a spectrophotometer. One unit of SOD was defined as the amount of enzyme necessary to inhibit formation of blue formazan to reduce absorbance to 50% of blank's.

CAT activity was measured in accordance with Beers and Sizer ([1954]). One ml of 59 mM H_2O_2 (Merck's Supersol) and 1.9 ml of 50 mM potassium phosphate (pH 7.0) were put in the quartz cuvette. 0.1 ml of enzyme extract was added. The absorbance was measured at 240 nm using a spectrophotometer adjusted to 25°C. Finally, decrease in absorbance for 3 min was recorded. One unit decomposes one micromole of H_2O_2 per minute at 25°C and pH 7.0. The catalase activity was obtained by the following equation (Beers and Sizer [1954]):

$$\text{Units/mg protein} = (\Delta A240/\text{min} \times 1000) /(43.6 \times \text{mg enzyme} /\text{ml reaction mixture}) \tag{3}$$

DPPH free radical scavenging activity was measured according to Lee et al. ([2008]) based on Blois method ([1958]). A total of 1 g of milled seed sample was extracted by 10 ml methanol at 50°C for 24 hours on a 150 rpm shaker. The filtered sample solution was centrifuged at 12,000× *g* for 20 min, and the supernatant was used for the assay. The reaction mixture containing 1,700 µl of 0.15 mM DPPH in methanol and 300 µl of sample solution was shaken well and incubated for 30 min. at room temperature, and the absorbance of the resulting solution was read at 517 nm. The scavenging activity was determined by comparing the absorbance with that of the control containing equal volumes of DPPH solution and ethanol. The DPPH radical scavenging activity was obtained by the following equation:

$$\text{DPPH radical scavenging activity}(\%) = [(A_{control} - A_{sample})/A_{control}] \times 100) \tag{4}$$

5.5 Statistical analysis

For the evaluation of individual treatment means, all collected data from the complete randomized design with three replicates (four replicates in germination test, exceptively), subjected to analysis of variance using Statistical Analysis System (SAS 9.1).

ACKNOWLEDGMENTS

This work was supported by a National Research Foundation of Korea Grant funded by the Korean Government (2011–0003554).

AUTHORS' CONTRIBUTIONS

JSB carried out seed freezing test, seed viability and physiochemical test under accelerated aging treatment, and wrote draft of manuscript. NJC designed research, carried out plant material selection, seed wintering test, and wrote manuscript. All authors read and approved the final manuscript.

REFERENCES

1. AOAC: *AOAC methods, official method of analysis of the official analytical chemists*. Association of Official Analytical Chemists, Washington; 1970.

2. AOAC: *AOAC methods, official method of analysis of the official analytical chemists*. Association of Official Analytical Chemists, Washington; 1990.

3. Baek H, Kim H, Yoon M, Park J, Choi W: **Effects of dry heat treatment on cryopreservation of onion seed.***Korean J Intl Agri* 2000, **12:** 97–107.

4. Bailly C: **Active oxygen species and antioxidants in seed biology.** *Seed Sci Res* 2004, **14:** 93–107.

5. Bandoniene D, Murkovic M: **The detection of radical scavenging compounds in crude extract of borage (Borago officinalis L.) by using an on-line HPLC-DPPH method.** *J Biochem Biophys Methods* 2002, **53:** 45–49.

6. Beers RF, Sizer IW: **Sulfide inhibition of catalase.** *Science* 1954, **120:** 32–33.

7. Bewley JD, Black M: *Physiology and biochemistry of seeds: in relation to germination, Vol. 2. Viability, dormancy and environmental control*. Springer- Verlag, Berin New York; 1982.

8. Beyer W: **Assaying for superoxide dismutase activity: some large consequences of minor changes in conditions.** *Anal Biochem* 1987, **161:** 559–566.

9. Blois MS: **Antioxidant determinations by the use of a stable free radical.** *Nature* 1958, **181:** 1199–1200.

10. Bowler C, Montagu M, Inze D: **Superoxide dismutase and stress tolerance.** *Annu Rev Plant Biol* 1992, **43:**83–116.

11. Cakmak I, Horst W: **Effect of aluminium on lipid peroxidation, superoxide dismutase, catalase, and peroxidase activities in root tips of soybean (Glycine max).** *Physiol Plant* 1991, **83:** 463–468.

12. Cao QJIN, Li B, Song ZHIP, Cai XX, Lu BAOR: **Impact of weedy rice populations on the growth and yield of direct seeded and transplanted rice.** *Weed Biol Manage* 2007, **7:** 97–104.

13. Chung N-J: **Elongation habit of mesocotyls and coleoptiles in weedy rice with high emergence ability in direct-seeding on dry paddy fields.** *Crop Pasture Sci* 2010, **61:** 911–917.

14. Chung N-J, Ahn S-N: **Genetic diversity and classification of Korean weedy rice based on RAPD markers.***Korean J Breed* 2005, **37:** 162–168.

15. Chung N-J, Paek N-C: **Photoblastism and ecophysiology of seed germination in weedy rice.** *Agron J*2003, **95:** 184–190.

16. Chung N, Yoon Y, Kim C, Kang Y: **Weedy rice control by no-tillage direct seeding on flooded paddy field.***Korean J Crop Sci* 2000, **45:** 195–198.

17. Cohn M, Hughes J: **Seed dormancy in red rice (Oryza sativa) I. Effect of temperature on dry-afterripening.***Weed Sci* 1981, **29:** 402–404.

18. Delouche JC, Burgos NR, Gealy DR, de San MGZ, Labrada R, Larinde M, Rosell C: *Weedy rices: origin, biology, ecology and control. FAO plant production and protection paper 188*. Food and Agriculture Organization of the United Nations, Rome; 2007.

19. DeVries A: **Antifreeze glycopeptides and peptides: interactions with ice and water.** *Methods Enzymol*1986, **127:** 293–303.

20. Fogliatto S, Vidotto F, Ferrero A: **Effects of winter flooding on weedy rice (Oryza sativa L.).** *Crop Prot* 2010,**29:** 1232–1240.

21. Fogliatto S, Vidotto F, Ferrero A: **Germination of weedy rice (Oryza sativa) in response to field conditions during winter.** *Weed Technol* 2011, **25:** 252–261.

22. Fujikura Y, Karssen CM: **Molecular studies on osmoprimed seeds of cauliflower: a partial amino acid sequence of a vigour-related protein and osmopriming-enhanced expression of putative aspartic protease.** *Seed Sci Res* 1995, **5:** 177–181.

23. Gu X, Chen Z, Foley M: **Inheritance of seed dormancy in weedy rice.** *Crop Sci* 2003, **43:** 835–843.

24. Hsua CC, Chena CL, Chenb JJ, Sunga JM: **Accelerated aging-enhanced lipid peroxidation in bitter gourd seeds and effects of priming and hot water soaking treatments.** *Sci Hortic* 2003, **98:** 201–212.

25. ISTA: *International rules for seed testing*. International Seed Testing Association, Basserdorf, Switzerland; 2009.

26. Ito O, Ella E, Kawano N: **Physiological basis of submergence tolerance in rainfed lowland rice ecosystem.** *Field Crop Res* 1999, **64:** 75–90.

27. Jeong I-H: *Identification of antifreeze proteins and freezing-resistance genes expressed during cold acclimation in wheat chromosome substitution lines*. Sunchon University, Dissertation; 2009.

28. Justice OL, Bass LN: *Principles and practices of seed storage. Agriculture Handbook No. 506*. USDA, Washington; 1978.

29. Kane NC, Baack EJ: **The origins of weedy rice.** *Mol Ecol* 2007, **16:** 4423–4425.

30. Kibinza S, Bazin J, Bailly C, Farrant JM, Corbineau F, El-Maarouf-Bouteau H: **Catalase is a key enzyme in seed recovery from ageing during priming.** *Plant Sci* 2011, **181:** 309–315.

31. Kim DH: **Native tees seeding of cryopreservation by Korea.** *National Forestry Cooperatives Federation Korea* 2010, **10:** 92–95.

32. Kim M, Lee S: **Resistance functions of woody landscape plants to air pollutants, 1; SOD activity.** *J Korean For Soc* 1992, **81:** 164–176.

33. Ko BR: *Studies on germination and growth characteristics of weedy rice(Oryza sativa L.) occurring in direct seeding cultivation on paddy field*. Wonkwang University, Dissertation; 2001.

34. Larson R: **Plant defenses against oxidative stress.** *Insect Biochem Physiol* 1995, **29:** 175–186.

35. Lee K, Kim Y, Chung N: **Determination and isolation of antioxidative serotonin derivatives, N-(p-Coumaroyl) serotonin and N-feruoylserotonin from safflower seeds.** *Korean J Crop Sci* 2008, **53:** 171–178.

36. McDonald MB: **Seed deterioration: physiology, repair and assessment.** *Seed Sci Technol* 1992, **27:** 177–237.

37. Nelms C, Twedt D: **Seed deterioration in flooded agricultural fields during winter.** *Wildl Soc Bull* 1996, **24:**85–88.

38. Noldin J, Chandler J, McCauley G: **Seed longevity of red rice ecotypes buried in soil.** *Planta Daninha* 2006,**24:** 611–620.

39. Ottis B, Smith K, Scott R, Talbert R: **Rice yield and quality as affected by cultivar and red rice (Oryza sativa) density.** *Weed Sci* 2005, **53:** 499–504.

40. Pantone D, Baker J: **Weed-crop competition models and response-surface analysis of red rice competition in cultivated rice: a review.** *Crop Sci* 1991, **31:** 1105–1110.

41. Pantone D, Baker J, Jordan P: **Path analysis of red rice (Oryza sativa L.) competition with cultivated rice.** *Weed Sci* 1992, **40:** 313–319.

42. Pilar R: **Lipoxygenase activity and lipid composition of cotyledons and oil bodies of two sunflower hybrids.** *Plant Physiol Biochem* 1998, **36:** 285–291.

43. Roberts EH: **Storage environment and the control of viability.** In *Viability of Seeds*. Edited by: Roberts EH. Chapman and Hall, London; 1972:14–58.

44. Savary S, Castilla N, Elazegui F, Teng P: **Multiple effects of two drivers of agricultural change, labour shortage and water scarcity, on rice pest profiles in tropical Asia.** *Field Crop Res* 2005, **91:** 263–271.

45. Scandalios J: **Oxygen stress and superoxide dismutases.** *Plant Physiol* 1993, **101:** 7–12.

46. Schoner S, Heinrich Krause G: **Protective systems against active oxygen species in spinach: response to cold acclimation in excess light.** *Planta* 1990, **180:** 383–389.

47. Seong K, Ku Y, Rho S, Han H: **Control of weedy rice in paddy field of rice.** *Treat. Of Crop Res* 2004, **5:** 507–511.

48. Shewfelt R, Purvis A: **Toward a comprehensive model for lipid peroxidation in plant tissue disorders.** *HortSci* 1995, **30:** 213–218.

49. Shon J, Ko J, Kim W, Kom B, Kim C, Jung N: **Changes of antioxidative enzymes and alcohol dehydrogenase in young rice seedlings submerged in water.** *Korean J Crop Sci* 2008, **53:** 440–446.

50. Suh H: **Genetic characterization and utilization of weedy rice.** *Korean J Breed Soc* 2003, **35:** 1–11.

51. Suh H: *Weedy rice*. Wild Crop Germplasm Center of Yeungnam University, Korea; 2008.

52. Tabbal D, Bouman B, Bhuiyan S, Sibayan E, Sattar M: **On-farm strategies for reducing water input in irrigated rice; case studies in the Philippines.** *Agric Water Manag* 2002, **56:** 93–112.

53. Tanida M: **Catalase activity of rice seed embryo and its relation to germination rate at a low temperature.***Breed Sci* 1996, **46:** 23–27.

54. Tomita S, Miyagawa S, Kono Y, Noichana C, Inamura T, Nagata Y, Sributta A, Nawata E: **Rice yield losses by competition with weeds in rainfed paddy fields in north east thailand.** *Weed Biol Manage* 2003, **3:** 162–171.

55. Vartapetian B, Jackson M: **Plant adaptations to anaerobic stress.** *Ann Bot* 1997, **79:** 3–20.

56. Vaughan KL, Ottis B, Prazak-Havey A, Bormans C, Sneller C, Chandler J, Park W: **Is all red rice found in commercial rice really Oryza sativa?** *Weed Sci* 2001, **49:** 468–476.

57. Yan B, Dai Q, Liu X, Huang S, Wang Z: **Flooding-induced membrane damage, lipid oxidation and activated oxygen generation in corn leaves.** *Plant Soil* 1996, **179:** 261–268.

58. Yoo J-Y, Hwang CH: *Development of the freeze-storage technology for vegetables using an AFP gene from cool-season turt grass*. Dankook University, Dissertation; 2000.

Chapter 8

Chat (Catha edulis): a socio economic crop in Harar Region, Eastern Ethiopia

Laxman S Kandari, Hiranmai R Yadav, Ashok K Thakur and Tripti Kandari

ABSTRACT

Chat (*Catha edulis*) is an important perennial crop and its leaves are chewed for a stimulating effect. It is widely cultivated in the Ethiopian highlands of Oromia region and is figured as Ethiopia's second largest foreign exchange earner. Its cultivation accounts for about 70% of farmer's income in the study area. The common effect of its consumption leads to insomnia, a condition that the users sometimes try to overcome with sedatives or alcohol. The present study is an attempt to survey and assess the impact of crop on the community. It has been observed to implicate health problems, reduces savings and nutritional standards of the family members. The chat yields in the area ranges from 1500–1800 kg/ha through monoculture. During the study, the average monthly income of the family practicing chat cultivation was from Birr 8, 533.00 to 13, 166.00 kg/ha per year in Baate and Genede cultivating areas. When the average cost per/ha was rupees 60/kg. The present study shows that during the recent past, leaf consumption has increased significantly. Chat growers are not only producers but also traders and consumers. Its consumption has become a widespread habit from secondary schools. Highest number of consumers was found to be among drivers followed by students and shopkeepers. The consumption of the plant is not considered a taboo but on contrary a status symbol in the region. It has no legal or moral implications and is considered as a part of custom and habit of local people. High value cash crop like vegetables and orchard fruits needs to be used as a replacement for chat which could be a regular source of income to farmers. Alternative sources of income for farmers needs to be scientifically worked out and proposed keeping in view the proportion of agricultural land reserved under chat cultivation and to increase the production of food grains being produced.

KEYWORDS

Agriculture, *Catha edulis*, Oromiya, Socio-economic impact

1. INTRODUCTION

Chat (*Catha edulis* Forsk), or Quat belonging to family Celastraceae is considered an evergreen plant, cultivated for the production of leaves having sympathomimetic actions which are used commonly for gradual chewing (Cox and Rampees 2003). The bole is straight and cylinder. The bark is whitish and crown is small. Flowers are small and white in colour with five petals and five stamens in axillary cymes 1–8 cm long. The calyx is five lobed; the capsule is oblong, woody, pendulous about 1 cm long with 3 valves. This plant is called by different names in different countries: 'chat' in Ethiopia, 'qat' in Yemen, 'mirra' in Kenya and 'qaad' or 'jaad' in Somalia. In most of the literature, it is largely known and called as chat (Belwal and Teshome 2011). In Ethiopia, this is grown extensively in the middle altitudes between 1500 and 2100 meters above sea level (masl), and performs better on well-drained soil under diverse climatic conditions. It can tolerate drought conditions for several months. The crop can be harvested around the year, thereby becoming a source of continuous revenue for the farmer. The economically important parts of the plant are its young leaves and tender twigs, which are chewed for their stimulating effect. It is not uncommon to come across many farm ladies selling their plants to willing buyers in the local market throughout the day (Figure 1). Chat production and consumption occupy a major area in eastern Africa, South-west Arabia, and Madagascar (Pantelis et al. 1989). Its consumption can alleviate fatigue, promote excitation, increase confidence and suppress sleep and hunger (Al-kamel 2001).

Figure 1. Woman chat seller selling chat to local people.

Ethiopia is the world's largest producer of chat which has recently become the fastest growing export commodity. The history of domestication and introduction of this crop in Ethiopia is not known. According to the folklore, it was first introduced in Harar from where it spread to rest of the country (Getahun and Krikorian 1973). Harar is located in eastern Ethiopia. The main inhabitants of this region are Oromia tribal communities. About a third of the production is exported to neighboring countries like Djibouti and Somalia. However, it is largely produced, marketed and consumed within the country. Harar is a business centre where the plant is grown at a wide range of altitudes, soil types and climatic conditions (Alles et al.1961; Beguinot 1939; Beitter 1900; Brooke 1960). Cultivation of this crop has helped to a good extent to reduce poverty from the region. The price of chat fluctuates seasonally according to the quality and quantity supplied. The price is higher in the dry periods due to short supply but decreases during the rainy season (from April to September), when the supply is more. Currently, chat crop is grown in about 94,330 ha of land nationwide and accounts for about one third of the area under coffee (Anonymous 2000). The bulk of the chat produced in Harar region is of good quality and is in great demand in both domestic as well as in export markets. It is now a major source of foreign exchange and occupies second place in export after coffee. In 1998–99, the crop accounted for 13.4% of Ethiopia's export earnings and was the country's second largest export items that year (CSA 2000). Consumption of chat leaves is common in Middle-East Asian countries like Yemen, Saudi Arabia and East African countries like, Kenya, Somalia, Djibouti, Uganda and Tanzania. Normally chat saplings are first planted around the homestead and later expanded to rest of the farmland. Cultivation of chat occupies 55% of agricultural land and is largely intercropped either with maize or with sorghum (Figure 2).

Figure 2. Intercropping of chat with Sorghum.

Many users report that feeling of happiness associated with better thinking capacity. Currently, majority of the farmers are involved in chat cultivation in this region. It is widely enjoyed by both male and females and is commonly used for social recreation and family ceremonies. Occupational groups such as motor vehicle and truck drivers chew plant leaves regularly during long distance driving and admit that it keeps them alert and awake. A significant number of students have also been observed to consume the leaves to remain alert especially during examination periods. There is also a specific usage of this plant by the special sections of the community like craftsmen and workers in small scale industries use it to reduce physical fatigue and traditional healers to heal ailments. The plant is interpreted as a stimulant of physical and psychological functions. Its psychic influence depends on its active ingredients that have a stimulating and cuphoric effect. The plant is chewed for a stimulating effect that is similar to that of amphetamine. Cathinone, found in fresh leaves is listed as a schedule I drug in the United States in the same group as heroin and cocaine (Brooks 1997). During maturation and decomposition of the plant, cathinone is converted to cathine, a Schedule IV drug (legal). In fact, for a variety of reasons, the habit of chat consumption is deeply rooted in social and cultural traditions of Yemen, Ethiopia, Somalia, and Djibouti (Getachew 1996; Bali 1997; Klingele 1998). The present study is an attempt to survey and assess the impact of crop on the community.

2. MATERIALS AND METHODS

2.1 Study area

The present study area encompasses the Haramaya town and adjacent areas in eastern-Ethiopia (Figure 3). The information and findings presented are primarily based on the interviews through a prepared questionnaires held with the traders and clients in the Baate and Gendaje areas of Harar and other consumers belonging to all categories of people i.e., local people, drivers, shopkeepers and students. The trading of chat is highly pronounced in the Harar region. Although people are subjected to modern facilities, the multi-ethnic composition of the population still clings to traditional beliefs and practices that are deeply rooted in their mind and culture.

2.2 Selection of informants

A reconnaissance survey was undertaken during 2011–2012 in the market and nearby adjoining area of Harar region i.e., Baate and Gendaje and where the plant is commonly sold. During the survey, attempts were made to collect all possible information regarding the traditional and ethnobotanical use of plant in the region; mode of usage and part of plant used. A semi-structured questionnaire survey with focused group discussion (FGD) were conducted with different groups of people i.e., among all categories of people. Most of the local people identified during the field survey were in the age category of 20–40 and > 50 years who were habitual and willing consumers. All of them were familiar with the use of the plant for cultivation and its subsequent consumption pattern.

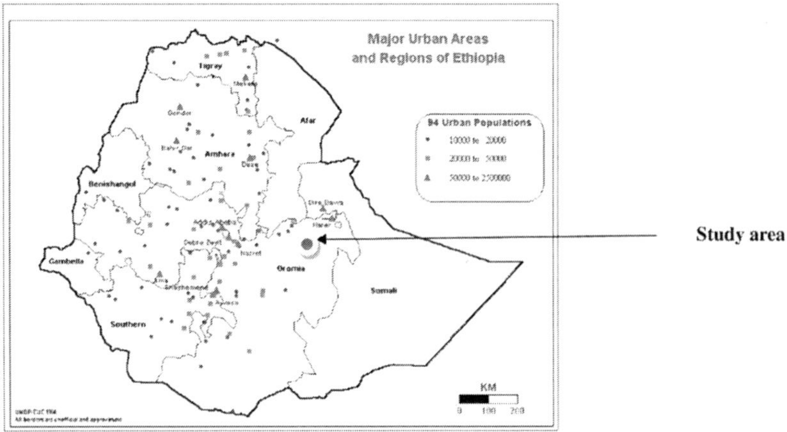

Map source : UNDP

Figure 3. Map of the study site.

3. RESULTS AND DISCUSSION

3.1 Botany and phenology of the *Catha edulis* plant

Catha edulis is a flowering evergreen large shrub of Celastraceae family. The plant usually grows at high altitude areas (1500–2500 m asl) in Ethiopia, Kenya, Yemen, South Africa and Madagascar. The plant reaches up to 25 m in height when grown naturally (Klingele 1998), but it is kept under manageable height when grown as a cash crop to facilitate convenient height for harvest and good economic return (Figure 4).

Figure 4. A full mature chat plant.

3.2 Socio-economic effect and expansion of chat production

It is observed that chat growers are not only producers but also traders and consumers. Its consumption has become a widespread habit from secondary schools. Highest numbers of consumers were found to be among drivers followed by students and shopkeepers (Figure 5). The leaves intake in general, negatively affects the working ability and capacity of the users. In fact, its consumers have negative health effect with positive income effect. As, such, policy measures required to control its cultivation will need to address both supply and demand issues in relation to its negative impact on the society. Chewing of leaves has been practiced for social and psychological reasons for many centuries and its use has gradually expanded worldwide. Both socioeconomic and agro-ecological reasons have contributed for its expansion and production thereby increasing market opportunities (Figure 6). Production of the plant is mainly located close to the road network and on farms with irrigation facilities. It has assumed the status of cash crop with an assured income. The profitability of its production is, therefore, considered as the primary reason by 75% of the farmers for expansion of cultivation in the area. The income generation from intercropping with maize is another added advantage of high returns as compared to mono-cropping of agriculture. Income from maize mono-cropping is less as compared to intercropping system, which provides 3 times higher return. Due to assured and higher returns from cultivation of the plant as well as low income from other crops or mono-crops, many farmers (having small landholdings) have expanded its cultivation in their fields. The other factors contributing to the expansion of this crop includes low risk and low labor intensive compared to cereal crops. Ownership of chat area under cultivation is considered as one of the important indicators of prosperity when a man proposes marriage to a woman in this region. The production of this crop has expanded at the expense of important rainfed cereal crops like maize, sorghum, coffee etc. and serves as a good substitute, as it is less vulnerable to drought. The cultivation of cereals is expensive, as it requires fertilizers and irrigation. The crop being perennial, farmers in the area considered this as a way to ensure land entitlement because annual cropland is more affected by land redistribution than land under perennial crops.

Agricultural labours also show more interest to work in chat cultivated fields as its cultivation offers wage or bonus to the workers. Credit is also available more easily for chat growers as the income is more or less assured. It has become the major source of income in the region and constitutes approximately 70% of a farmer's income. It was found that nearly 3.5–4.28 numbers of family members out of 5 were involved in cultivation of this plant in the area Gendaje and Baate (Table 1). The chat yields in the area ranges from 1500–1800 kg/ha through monoculture. During the study, the average monthly income of the family practicing chat cultivation was from Birr 8, 533.00 to 13, 166.00 kg/ha per year in Baate and Genede cultivating areas. When the average cost per/ha was rupees 60/kg. The present study shows that during the recent past, leaf consumption has increased significantly. Its cultivation continuous to play an important role in the social life of people in most parts of Ethiopia particularly in celebrations, marriages, business and political meetings (Figure 5). A survey of the Harar region revealed that, the use of plant leaves has become popular among all segments of the population and different age groups (Selassie and Gebre 1996). Its use was previously limited to its producing areas, which has now spread to all parts of the country. It is now air freighted and available in Europe, Canada and the United States, following the migration routes of immigrants from east Africa and southern Arabia.

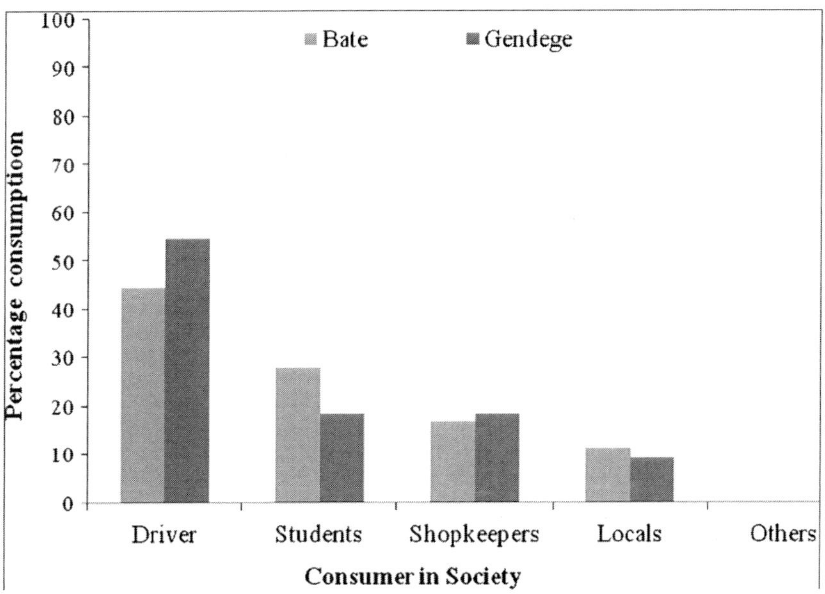

Figure 5. Consumption of chat by different sector of society.

Specific preference for cultivation of this plant has resulted in very glaring agriculture scenario in the area

1 Provision of cash crop which provides assured income to the farmers throughout the year at regular intervals.

2 Increase in goat population which could be raised with the little inputs that brings in revenue through meat sales during festive periods.

3 Raising employment opportunities through domestic and international trade along with increased transportation.

4 Easy availability of labour.

5 Negligible need for irregular and vital inputs.

6 Easy credit availability.

7 It enjoys a status of important foreign exchange earner.

8 The crop assured elevated social status for the producers.

Figure 6. Local people enjoying chat. (a) **Enjoying chat in home.** (b) **A labour enjoying chat in free working hours.**

3.3 Impact and response of society and consequences

Development of withdrawal symptoms, comprising heavy and sinking sensatation following habit of prolonged consumption of chat seems to surfacing among consumers. The frequency of lethargy, mild depression, slight trembling and recurrent bad dreams prompt them to have second thoughts about its consumption. A serious consideration is that, its use may endanger health, thereby resulting in anorexia leading to malnutrition with subsequent susceptibility to infectious diseases. As a result of area expansion under this plant, the farmers need to buy cereals now to meet their food requirements and thus its production has resulted in problem of food insecurity in the region. The farmers' personal production of food crops may last for 6–7 months only. Only about 51% of the households have enough food throughout the year. Chat consumption negatively affects the working capacity of people because they tend to be slow in work, show lethargy, less number of working hours, take frequent rests, spend time chewing the leaves, and are generally more careless and found loitering about aimlessly in the market. Its consumption has become a common habit among the population above 13 years of age including the secondary school students. Consumers spend a high portion of their income to purchase these leaves which leads to a serious under nourishment of the poor people leading to social consequences. The number of goat population has increased in the current farming system due to chat foliage availability and less dependence on crop residues for feed along with different fodder requirements. The labour requirement for chat cultivation is lower than cereal mono-cropping with total volume of employment rising due to ancillary post harvest activities, trade and transportation. In Somalia, it is reported that the plant consumption is promoting different types of criminal activities (Elmi et al. 1987).

Its increased popularity however has brought in important social problems in its wake

1 Health problem like mild depression and trembling

2 Malnutrition and susceptibility to infectious diseases

3 Low food grain production resulting in increased dependence on market for food grains.

4 Affected chewing capacity of people accompanied with lethargy

5 Affected school children resulting in school dropouts

6 Increase in criminal activities

7 Decrease in area of cultivation under cereal crops with subsequent drop in production of food grains.

8 Due to fall in area under fodder crops the livestock population along with milk production has been on a downswing

Table 1. Socio-economic dependency of people on chat

S. No.	Parameters	Villages	
		Genede	Baate
1.	Number of families members involved in chat business	3.5 ± 0.88	4.28 ± 0.821
2.	Number of families have their own land	33.3 ± 0.166	21.42 ± 0.10
3.	Working hrs per day	14.8 ± 0.77	14.4 ± 0.77
4.	Monthly income	131, 66.00 ± 5296	8,533.00 ± 890.21
5.	Price of one bunch of chat	58 ± 9.27	41.32 ± 6.412

±Standard error.

The only positive effect of chat production is higher income generation for farmers, more employment generation to the rural people engaged in chat transportation and trade related activities.

3.4 Chat expansion on the farming system

It has been observed that production of this crop in the study area has expanded to almost 55% of the cultivated area in Harar which includes intercropping with maize and (to some extent with) sorghum. In the intercropping category, this plant has been found to have the largest share of land resulting in decreases in yield of maize and total cereal production in the area. The expansion of area under chat production has also changed the composition of livestock system in the region due to decreased fodder availability and less interest in livestock raising. The need for drought animals like oxen has reduced in the ongoing pattern of chat-based cultivation system. Further, fodder availability has been reduced because chat occupies the larger share of the land coupled with low animal population. Fodder species like *Erythrina abyssinica*, *Sesbania sesban* and *Leucaena leucocephala* are completely replaced or removed. Shifts in the cropping pattern also resulted in decreased number of cows with decreasing milch cattle and the milk availability in the market is currently on the downswing.

3.5 Intercropping

In the area surveyed rainfed farming is undertaken and farmers need food crop for their annual domestic consumption needs for which Sorghum and sweet potato is cultivated. Maize is sometimes also cultivated instead of Sorghum depending upon seed availability. Sorghum, corn and sweet potatoes are generally accepted for intercropping with chat which is a cash crop. This is possible as these plants are so spaced that they leave much area in between the rows (Figure 2). The preparation of the seedbed required for intercropping with Sorghum provides higher and additional benefits. Thus, the intercropping is practiced to improve the economy of the farmers. On the other, chat leaves could be collected and marketed at regular intervals unlike most agricultural crops that are seasonal which is a welcome source of the otherwise poor cultivation. Vegetables crops are not preferred in this region and are cultivated where water is available.

3.6 Biochemical constituent

The active ingredient of chat responsible for its psycho stimulant effect is an alkaloid chemical known as cathionine, which is structurally and chemically similar to d-amphetamine and cathine a milder form of cathionine. Cathionine is a highly potent stimulant, which produces sympathomimetic and central nervous system stimulation analogous to the effect of amphetamine. Fresh leaves contain both ingredients: those left unrefrigerated beyond 48 hours would contain only cathine, and it supports user's preference for fresh leaves. The plant loses its potency after 48 hours. The results of various *in-vivo* and *in-vitro* experiments indicated that the substance could be considered as a natural amphetamine. The active constituent, cathinone, which causes sympatno-mimetic effects and symptoms such as euphoria and hyperactivity, is released with following consumption of leaves.

Cathinone has analogous mechanisms of action with pharmacological properties that are reminiscent of those induced by amphetamine, i.e. anorexia as well as hypermotility (Valterio and Kalix 1982). In fact, it is now being referred to as a "natural amphetamine" and its effects in animals correspond to those observed in human consumers (Alem et al. 1999). The World Health Organization (WHO) has included and reported cathinone in its list of controlled drugs and equivalent to amphetamine abuse (WHO 1980; Kalix 1981). The plant leaves are considered rich in ascorbic acid, which minimizes the undesirable side-effects of its chewing. By modulating catecholaminergic activity or transmission of dopamine, ascorbic acid acts as an antidote to the effects of amphetamine (Broody 2002; Gulley and Rebec 1999). It is assumed that there is an adequate quantity of ascorbic acid in the leaves, which is released from the leaf matrix by mastication and becomes bio-available. Ascorbic acid, a well-known antidote against amphetamine and amphetamine-like substances, was reported to reach as much as 325 mg/100 g leaf of chat (Krikorian and Getahun 1973).

In light of World Health Organization (WHO) recommendation, the problems associated with chat-chewing for the moment should be considered in a manner similar to amphetamine abuse. Hence, it is the psychotropic and mind-altering drug type whose use could possibly constitute risk behavior in the amplification of the HIV/AIDS epidemic in countries like Ethiopia, where the habit is widespread (Abebe et al. 2005).

Keeping in view the adverse side effect of consuming chat at fall out on the health of younger generation of the region, it is imperative that crucial step be undertaken for production and consumption of this plant.

RECOMMENDATIONS

Since, impediments in its cultivation would affect the economy of the area, the following point need to be considered for implementation.

1 Legislation need to be enforced for confining its cultivation to a particular region.

2 Government machinery needs to be in place for collection of harvest at market support price and for meeting exports.

3 Proper guidelines should be placed for its consumption. Younger generation should be prohibited from consuming it

4 Providing education for the students to stay away from its consumption by highlighting its side effects.

5 Establishing government laboratories for analysing the plant and giving the task of enforcing ban on its consumption by younger generations.

6 Establishing chat free areas in the country.

7 Creating revenue earning demonstration areas of vegetable and flower cultivation to wean away the farmers from the chat.

In Ethiopia, farmers have few alternative sources of income with similar profitability. Control on chat will therefore have to address both supply and demand. Alternative sources of income for farmers need to be designed and thoroughly investigated while providing other appropriate options. Education and awareness programs about consumption and its negative consequences may reduce the demand for chat.

CONCLUSION

From the current study, it is concluded that, chat production is expanding rapidly around Harar region, particularly at the expense of important cereal crops. The economic benefits and easy cultivation of the crop is a strong compelling reason for its expansion and increasing the income nearly 70% of farmers of the region. The crop being perennial, reduces the frequency of ploughing i.e., once in three months or yearly. Therefore, farmers are interested to cultivate it instead of food crops and to have more goats instead of oxen. Goats need less fodder and are more income generating. Chat producers become its perpetual users and its consumption is widespread among locals (both male and female) and even students. The expansion of this crop production in Ethiopia may bring considerable social and economic risks in the longer run as earnings totally depend as the plant is banned in many other countries. The consumption of this crop generates health problem and reduces the productivity in terms of work efficiency and number of working hours. The increased use of the plants needs to be monitored and controlled particularly among the younger generation who are being otherwise talked into imaginary well being. It is a evident from this study that the plant is very popular among all age group of people approached. The consumption is not common among people beyond 60 years of age. It has been supplementing the income of farmers in the absence of other cash crops. The crop now enjoys a well developed flourishing market which is instrumental in reducing poverty in the area.

ACKNOWLEDGEMENTS

Authors are thankful to Haramaya University for extending facilities for study and to local people and chat consumers of the region who constituted the essential components of the study.

AUTHORS' CONTRIBUTIONS

LSK collect the data and write the manuscript. HRY read and approved the final manuscript and help in data collection. AKT read and approved the final manuscript. TK collect the data help in writing and finally read and approved the final manuscript. All authors read and approved the final manuscript.

REFERENCES

1. Abebe D, Debella A, Dejene A, Degefa A, Abebe A, Urga K, Ketema L: Khat chewing habit as a possible risk behavior for HIV infection: a case-control study. *Ethiop J Health Dev* 2005, 3: 174-181.

2. Alem A, Kebede D, Kullgren G: The prevalence and socio-demographic correlates of khat chewing in Butajira, Ethiopia. *Acta Psychiat Scand* 1999, 100: 84-91.

3. Al-Kamel M: *Khat plant*. Ain-Shams Faculty of Medicine, Cairo, Egypt; 2001.

4. Alles GA, Fairchild MD, Jensen M: Chemical pharmacology of *Catha edulis* . *J Med Pharm Chem* 1961, 3: 323-352.

5. Anonymous: *US Department of Commerce. Ethiopia. Overall export down, "Khat" sales up*. 2000. Available at:http://www.tradeport.org Available at: (cited by Feylsa and Aune, 2003). WATH technical report No. 2., USA-AID West Africa Programme

6. Bali M: *Chewing khat: reflection on the Somali male food and social life*. 1997. Maroodi Jeex, Somaliland Alternative Newsletter. 6 Maroodi Jeex, Somaliland Alternative Newsletter. 6

7. Beguinot A: Notizie sulla *Catha edulis* Forsk. coltivata nel R. Orto Botanico di Genova. *Nuovo Giornale Botanico Italiano (NS)* 1939, 46: 332-334.

8. Beitter A: *Pharmacognostisch-chemische Untersuchung der Catha edulis. Inaugural -Dissertation der mathemat. und naturwissenschaftlichen Fakultat der Kaisers-Wilhelms-Universitat Strassburg. Strassburg i.E. Verlag von Schlesier & Schweikhardt*. 1900, 76.

9. Belwal R, Teshome H: Chat exports and Ethiopian economy: opportunities, dilemmas and constraints. *Afr J Bus Manage* 2011, 5(9):3635-3648.

10. Broody S: High-dose ascorbic acid increases intercourse frequency and improves mood: a randomised controlled clinical trial. *Biol Psychiatry* 2002, 52(4):371-374.

11. Brooke C: Khat (*Catha edulis*): its production and trade in the Middle East. *Geogr J* 1960, 126: 52-59. 10.2307/1790429

12. Brooks M: *Trade in Africa. TED case studies. Qat trade in Africa*. 1997. Available at: gurukul.ucc.american.edu/ted/QAT.HTM; accessed 22 May 2014.

13. Cox G, Rampees H: Adverse effect of Khat: a review. *Adv Psychiatr Treat* 2003, 9: 456-463.

14. CSA (Central Statistical Authority): *Statistical Abstract 1999*. Central Statistical Authority, Addis Ababa, Ethiopia; 2000.

15. Elmi AS, Ahmed YH, Samatar MS: Experience in the control of khat-chewing in Somalia. *Bull Narc* 1987, 39: 51-57.

16. Getachew M: *The cultivation of and uses of khat among the Oromo of Harar with particular reference to Haramaya Woreda*. Addis Ababa University, Addis Ababa, Ethiopia; 1996.

17. Getahun A, Krikorian AD: Chat: coffee's rival from Harar, Ethiopia. I. Botany, cultivation and use. *Econ Bot* 1973, 27: 353-377.

18. Gulley JM, Rebec GV: Modulatory effects of ascorbate, alone or with haloperidol, on a lever-release conditioned avoidance response task. *Pharmacol Biochem Behav* 1999, 63: 125-129.

19. Kalix P: Cathinone, an alkaloid from khat leaves with an amphetamine-like releasing effect.*Psychopharmacology* 1981, 74: 269-270.

20. Klingele R: *The Hararghie farmers on the crossroad between subsistence and cash economy. Field report*. UNDP–EUE, Addis Ababa, Ethiopia; 1998.

21. Krikorian AD, Getahun A: Chat: coffee's rival from Harar, Ethiopia. II. Chemical composition. *Econ Bot* 1973, 27: 378-389.

22. Pantelis C, Hindler CG, Taylor JC: Use and abuse of Khat (*Catha edulis*): a review of the distribution, pharmacology, side effects and a description of psychosis attributed to khat chewing. *Psychol Med* 1989, 19: 657-658.

23. Selassie SG, Gebre A: Rapid assessment of drug abuse in Ethiopia. *Bull Narc* 1996, 48(1-2):53-63.

24. Valterio C, Kalix P: The effect of the alkaloid (-) cathinone on motor activity in mice. *Arch Int Pharmacodyn Ther*1982, 255: 196-203.

25. WHO: Advisory group. Review of the pharmacology of khat. *Bull Narc* 1980, 33(3):83-93.

Chapter 9

Traditional agroecosystems as conservatories and incubators of cultivated plant varietal diversity: the case of fig (*Ficus carica*L.) in Morocco

Hafid Achtak,[1,2,3] Mohammed Ater,[3] Ahmed Oukabli,[4] Sylvain Santoni,[5] Finn Kjellberg[6] and Bouchaib Khadari[1,7*]

[1]INRA, UMR 1098, Développement et Amélioration des Plantes (DAP), Bat. 3, Campus CIRAD TA A 96/03, Av. Agropolis, 34398 Montpellier Cedex 5, France.
[2]Montpellier SupAgro, UMR 1098 DAP, Bat. 3, Campus CIRAD TA A 96/03, Av. Agropolis, 34398 Montpellier Cedex 5, France.
[3]Faculté des Sciences de Tétouan, Diversité et Conservation des Systèmes Biologiques, BP 2062, M'hannech II Tétouan, Maroc.
[4]INRA, UR Amélioration des Plantes et Conservation des Ressources Phytogénétiques, BP 578 Meknès, Maroc.
[5]INRA, UMR 1097, Diversité et Adaptation des Plantes Cultivées (DiA-PC), Bat. 33, 2 place Viala, 34060 Montpellier Cedex 2, France.
[6]CNRS, UMR 5175, Centre d'Ecologie Evolutive et Fonctionnelle (CEFE), 1919 route de Mende, 34293 Montpellier Cedex 5, France.
[7]Conservatoire Botanique National Méditerranéen de Porquerolles, UMR 1098 DAP, 76 A, Av. Gambetta, 83400 Hyères, France.

ABSTRACT

Background

Traditional agroecosystems are known to host both large crop species diversity and high within crop genetic diversity. In a context of global change, this diversity may be needed to feed the world. Are these agroecosystems museums (i.e. large core collections) or cradles of diversity? We investigated this question for a clonally propagated plant, fig (*Ficus carica*), within its native range, in Morocco, but as far away as possible from supposed centers of domestication.

Results

Fig varieties were locally numerous. They were found to be mainly highly local and corresponded to clones propagated vegetatively. Nevertheless these clones were often sufficiently old to have accumulated somatic mutations for selected traits (fig skin color) and at neutral loci (microsatellite markers). Further the pattern of spatial genetic structure was similar to the pattern expected in natural population for a mutation/drift/migration model at equilibrium, with homogeneous levels of local genetic diversity throughout Moroccan traditional agroecosystems.

Conclusions

We conclude that traditional agroecosystems constitue active incubators of varietal diversity even for clonally propagated crop species, and even when varieties correspond to clones that are often old. As only female fig is cultivated, wild fig and cultivated fig probably constitute a single evolutionary unit within these traditional agroecosystems. Core collections, however useful, are museums and hence cannot serve the same functions as traditional agroecosystems.

1. BACKGROUND

High yield agriculture based on elite crop varieties and high inputs results in loss of both numbers of crop plants and genetic resources within crops, thus threatening crop biodiversity and the nutritional safety of humanity [1]. To preserve crop diversity, traditional landscapes may have to be preserved [2]. In analogy with the concept of "biodiversity hotspot" used to identify priority areas for the conservation of wild species [3], traditional agroecosystems could be considered as main conservatories of crop biodiversity [4]. Indeed in 2002 the FAO started an initiative for the conservation and adaptive management of Globally Important Agricultural Heritage Systems http://www.fao.org/nr/giahs/en/. Although they are quite diverse, these agroecosystems exhibit common features such as 1) a high diversity of crop species, 2) the use of diversified traditional varieties, 3) sustainable agriculture, 4) low inputs associated with traditional farming practices and 5) the farmers obtaining a sizable proportion of their seeds (or vegetative equivalents) from their own harvest [5]. For instance, a survey in continental oases in northern Oman recorded 107 different crop species belonging to 39 families, including 33 fruit species [6]. This large biodiversity was successfully achieved despite the constraints of a small scale cropping system under arid and semi-arid conditions. Similarly, a study of 27 crop species in traditional agroecosystems distributed in eight countries over the five continents [7] demonstrated that such agroecosystems maintain considerable within crop genetic diversity. Traditional agroecosystems are either the repositories of crop diversity, or the place where extant crop diversity was fostered. Hence investigating within crop species genetic diversity and its geographic variation would help understanding genetic resources and dynamic processes of past and present domestication and subsequent diversification.

The biodiversity hotspot concept is associated with a major pattern of biodiversity: it increases close to the equator, and decreases towards the poles [8]. Two main ideas have been suggested to explain this global biodiversity pattern. Equatorial regions are a museum of biodiversity preserving ancient biodiversity, and/or they are a cradle generating new biodiversity [9]. If agroecosystems are hosting huge crop biodiversity, should we consider them as museums or as incubators of crop biodiversity, or as both? For long term crop management policies and hence to feed the world, the answers to this question is of a great importance.

The Mediterranean basin is one of 25 hotspots of biodiversity in the world. It hosts 25,000 species, of which 13,000 are endemic, this later group representing 4.3% of the worldwide flora [3]. It is the largest biodiversity hotspot on earth (over 2,000,000 km^2) and it includes several separate refuge areas [10]. Traditional agroecosystems are still found all over the Mediterranean region in mountains and oases. However several of these traditional agroecosystems may be of particular importance for preserving crop biodiversity. Indeed, many plant species were originally domesticated close to the Eastern shores of the Mediterranean. Hence, we might encounter contrasted patterns of genetic diversity within crops throughout the Mediterranean area, with more crop diversity available in the Eastern Mediterranean.

The process of domestication seems to have been diffuse, with prolonged cultivation of undomesticated forms, and prolonged genetic exchanges of domesticated forms with local wild relatives, at least for crops propagated by seeds [11, 12]. With a such domestication process, traditional agroecosystems located in the East Mediterranean may be most important for preservation of crop genetic resources. In addition, the domestication process of clonally propagated crops, particularly fruit trees, is often thought to have been an instant or almost instant process [13, 14] building on the idea that genotypes presenting the whole suite of agronomic traits of interest may have arisen by chance within totally natural populations [15]. This may qualify as a silver bullet hypothesis. If we follow this hypothesis, domestication was instantaneous, and followed by subsequent clonal propagation. Hence we would expect that extant varieties are old, probably limited in number, and that they represent the gene pool of the original region of domestication. The wild progenitors of some of these clonal crops still grow all around the Mediterranean region. This is true for three most symbolic crops in these regions such as olive, grape wine and fig. Therefore, we may ask, within such species, whether extant varietal genetic diversity in traditional agroecosystems reflects the propagation of old widespread clones, or old local clones, or recent local clones. We may even ask whether varieties could be fuzzy aggregations of genotypes (landraces) [16].

We chose to address this question in fig which presents us with a particularly fascinating situation as it is extremely easy propagated via cuttings, and was domesticated extremely early in the Near East, contemporarily with cereal crops, 9-12,000 BP [17]. Fig, *Ficus carica* L., is dioecious. Female trees produce the edible crop. Male trees produce pollen and their figs host the pollinator, *Blastophaga psenes* [18]. Each fig variety is a clone of female tree that are propagated through cuttings. Some fig varieties may produce seedless fig fruits without pollination while other varieties require pollination for successful fruit set [19]. Female figs produce seeds if pollinated. Male figs are often collected far from zones of fig cultivation and suspended in cultivated female trees to ensure pollination [20].

Phylogeographic studies based on cytoplasmic genes showed that wild fig was present all over the Mediterranean basin before domestication [21].

We investigated the genetic diversity of fig varieties in Moroccan traditional agroecosystems. Morocco is at the Western limit of the natural range of fig, as far away as possible (over 3500 km) from postulated places of domestication. Hence, if domestication begun and ended in the Eastern Mediterranean, then we expect to observe limited diversity so far away from the original zone of domestication. We also expect to observe lack of spatial genetic structure within Morocco, or simply a decrease of diversity when further away from the shores of the Mediterranean.

We made extensive collections of fig cultivars *in situ*, in order to 1) test whether cultivars are effectively highly local, 2) detect whether some of these cultivars are old, and 3) establish what insights into the history of fig cultivation could be drawn from extant genetic diversity and its spatial structuring.

We show here that in traditional agroecosystems, fig varieties are true clones, highly diversified, often highly local. Nevertheless they are often sufficiently old to have accumulated somatic mutations. Spatial genetic structure resembles what would be expected for a wild plant at mutation/drift/migration equilibrium. We conclude that the Moroccan traditional agroecosystems are at the same time museums and incubators of fig variety diversity, in a dynamic system preserving old, local varieties and generating new ones locally.

2. RESULTS AND DISCUSSION

277 cultivated trees were sampled throughout traditional Moroccan agroecosystems distributed over 40 sites that we grouped into 6 geographical zones (Figure 1). During field collection, we noted that, within each site, trees designated by the same name (local variety) shared highly similar morphological traits. To maximize genetic diversity of our sampling we generally collected a single individual per variety per site. Nevertheless, in a number of cases we sampled twice the same local variety within a site or within adjacent sites. Such samples systematically shared a same genotype. Hence genetic evidence confirms the obvious conclusion from phenotypic observation that local varieties are generally clones.

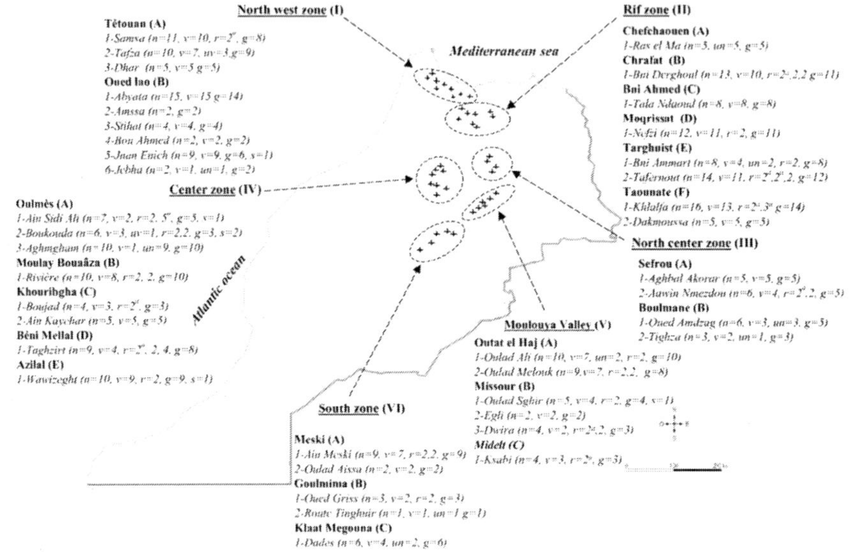

Figure 1. Sampling locations and fig sample diversity. Six geographic zones were defined I, II, III, IV, V and VI. Letters A, B, C, D, E and F correspond to subzones, and within each subzone, sites are indicated. v = number of sampled local variety names; un = number of sampled unnamed variety; for varieties sampled several times in a site, r is the number of repeats of each local variety name (r = 2, 3, 5 means that 3 varieties have been sampled several times, one 2 times, the second 3 times and the third 5 times); g = number of genotypes sampled; s = number of varieties presenting somatic mutations for fig skin color; a = fig trees under the same variety name and genotype.

2.1 SSR polymorphism and its discrimination power

The 277 individuals genotyped were separated into 194 distinct molecular profiles using 17 SSR loci (see Additional File 1). Genetic parameters for each locus are given in Table 1[22–25]. Overall, observed heterozygosity was higher than expected heterozygosity. The discriminating power per locus, D_i (probability of distinguishing two randomly chosen clones), ranged from 0.495 (LMFC26) to 0.979 (LMFC30) with a mean of 0.70 (Table 1). Hence the probability of confusing a randomly chosen clone with another one (under the hypothesis of statistical independence of the loci) was $\Pi(1-D_i) = 5 \times 10^{-11}$. With only 38,226 pairwise comparisons (including identical genotypes) in our data set, all cases of genotype identity should correspond to clones.

Table 1. Genetic parameters of the 17 SSR loci used in this study.

Locus	*A*	Size range (in bp)	H_O	H_E	F_{IS}	*D*
MFC1[a]	5	161-195	0.620	0.629	0.016	0.841
MFC2[a]	7	156-190	0.599	0.602	0.008	0.880
MFC3[a]	9	96-136	0.818	0.760	-0.074	0.851
MFC4[a]	5	216-226	0.524	0.493	-0.060	0.652
MFC8[b]	2	173-177	0.508	0.490	-0.033	0.619
MFC9[b]	7	188-211	0.636	0.582	-0.090	0.786
MFC11[b]	7	181-203	0.604	0.569	-0.059	0.585
MFC12[b]	4	152-167	0.578	0.552	-0.045	0.743
FSYC01[c]	5	117-160	0.455	0.451	-0.005	0.842
FSYC04[c]	2	181-183	0.529	0.502	-0.053	0.595
LMFC19[d]	8	296-312	0.433	0.398	-0.086	0.573
LMFC24[d]	4	272-278	0.460	0.456	-0.006	0.646
LMFC26[d]	3	224-236	0.235	0.223	-0.051	0.495
LMFC28[d]	5	192-203	0.562	0.558	-0.004	0.733
LMFC30[d]	11	231-261	0.904	0.820	-0.100	0.979
LMFC32[d]	9	197-225	0.433	0.415	-0.039	0.628
LMFC34[d]	2	245-247	0.492	0.486	-0.009	0.519

A: number of alleles, H_O: observed heterozygosity, H_E: expected heterozygosity, F_{IS}: within population fixation index, *D*: discriminating power. Primers developed by: [a] Khadari et *al.* [22], [b] Achtak et *al.* [23], [c] Ahmed et *al.* [24] and [d] Giraldo et *al.* [25].

We plotted the distribution of number of allelic differences between the 194 different genotypes in order to visualize the distribution of genetic differences between genotypes, (Figure 2A; 18,721 pairwise comparisons, excluding identical genotypes). The distribution ranged from 1 to 34 differences, presented a major peak at 19-20 differences and a very distinct, but very small, peak at 1-3 differences. The probability to observe by chance two or more genotypes that were distinguished by 3 alleles was 2.6×10^{-6}. Further, individuals whose genotypes were identical or differed by only 1-3 alleles were morphologically highly similar (see Additional File 2). The systematic association of genetic similitude for neutral markers with morphological similarity allows to conclude that all these trees belonged to a single original clone and that some had accumulated somatic mutations. Further, the shape of the pairwise genetic difference curve suggests that, beyond the case of the few genotypes deriving from each other by somatic mutations, all other genotypes are the product of sexual reproduction. We chose to be highly conservative in our estimate of which genotypes represented somatic mutations. Indeed the curve suggests that the limit may be better placed above 6 differences and indeed the probability of observing by chance two genotypes differing only by 6 alleles was still low, at 0.0017.

Hence, we classified the 194 genotypes into 152 genotype groups (clones) separated by at most 3 alleles, which were distinguished from all other genotypes by 4 to 34 alleles. Out of these groups of genotypes, 128 contained a single individual while 24 groups contained more than one individual and represented collectively 66 genotypes. Often a variety name was found to be associated with the same clone (identical or almost identical genotype) in different sites, conforting our conclusions. Numbers of trees sharing the same genotypes are given in Additional File 3, while Additional File 4and Figure 3 provide a series of cases of genotypes differentiated by 1-3 alleles and sharing the same variety name. While we have no data on mutation rates in somatic lines, the presence of such mutations within clonal lineages suggests that these varieties are old.

2.2 Variety names and characterization

Out of 277 sampled fig trees, 246 were named by the local farmers while for 31 fig trees, the interviewed farmers did not provide any name (see Additional File 1). These 31 unnamed trees corresponded to 30 genotypes out of which four corresponded genetically and morphologically to known varieties ('Ikoran Imelalen', 'El Messari', 'El kehla' and 'Beyota') and three were very similar genetically and morphologically to the 'Saaidi' and 'Rhoudane' varieties. The remaining 23 genotypes were distinct from previously defined varieties.

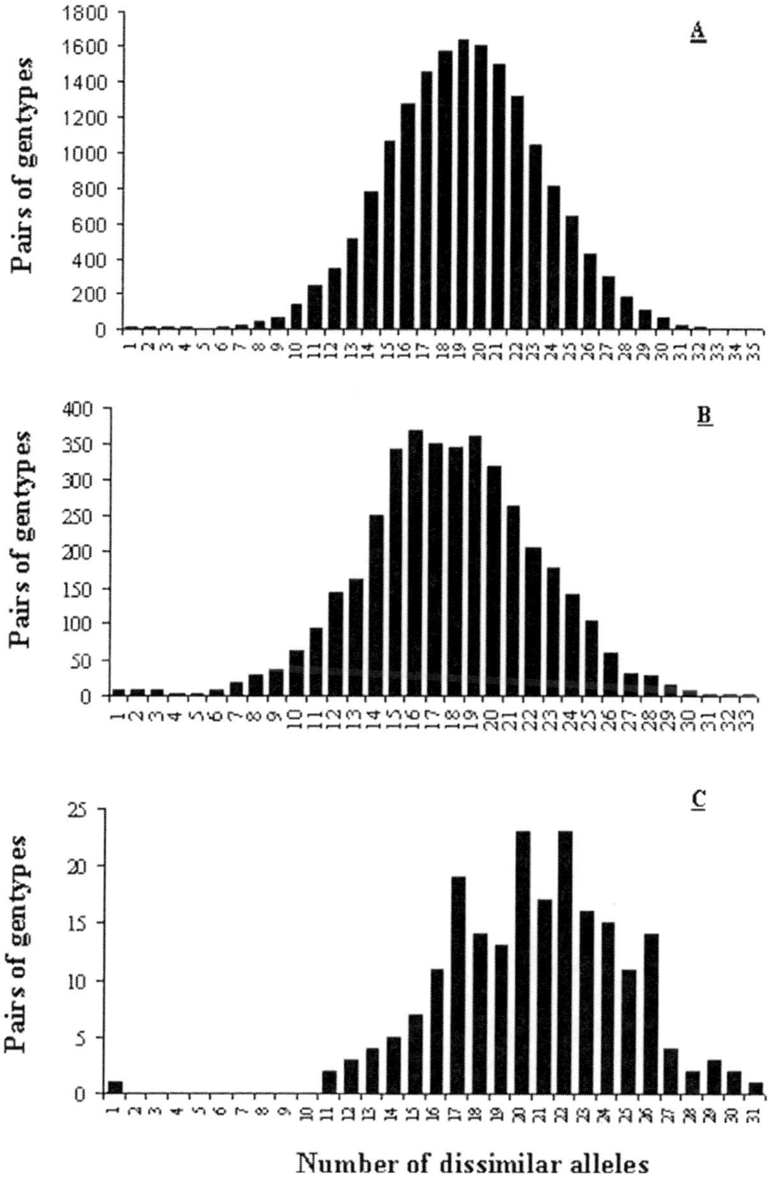

Figure 2. Frequency distribution of genetic dissimilarity for all pairwise comparisons between cultivated fig genotypes. (A) complete data set; (B) in mountain agroecosystem; (C) in oasis agroecosystem. Genetic differences among genotypes are retained in the oasis agroecosystem, despite the low number of genotypes cultivated (21). Note on the three graphs the bimodal shape of the curve with a very small peak for differences of 1-3 alleles.

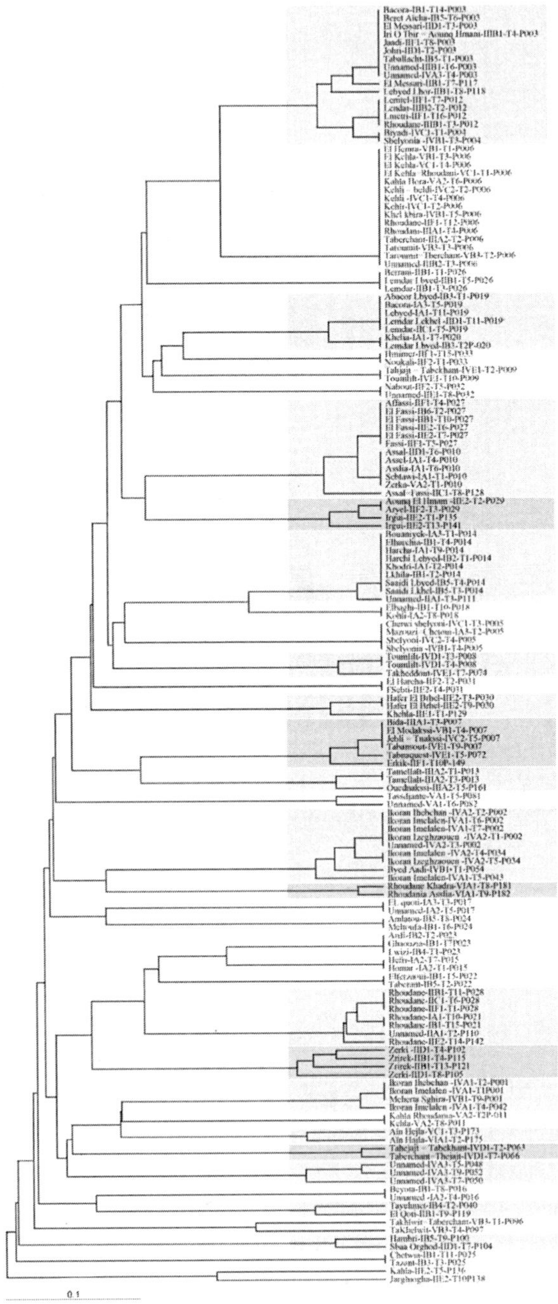

Figure 3. Genetic similitude among fig varieties. Samples grouped within a box correspond to highly similar genotypes that most probably derive from each other by somatic mutation. These similar genotypes often bear similar

variety names. After the variety name, the roman number indicates zone of sampling, the letter the subzone, followed by a number giving the precise site of sampling, Tx indicates the tree number × within site and Pxxx indicates genotype number xxx (see Additional File 1).

Synonymy was observed for 23 genotypes, with 2 to 7 denominations per genotype (Figure 3, Additional File 3). Two situations were observed. True synonymy was observed when the different fig trees presented identical pomological traits such as the varieties 'Johri' and 'El Messari' (green fig skin color, flattened pyriform fruit shape and red internal color). This situation was encountered for 20 genotypes. False synonymy was observed for fig trees known under the same generic denomination to which a descriptor of fig skin color was added. In these cases the leaves and the figs presented similar morphologies but fruit color was different. Six instances of the latter situation were encountered (Figure 3, Additional File 3). They included for instance 'Saaidi Lbyed-IB5-T4-P014' (white skin color) and 'Saaidi Lkhel-IB5-T3-P014' (black) in the North west zone, 'Ikoran Ihebchan-IVA2-T2-P002' (black) and 'Ikoran Imelalen-IVA1-T1-P001' (white) in the Center zone (Figures 1 and 3).

We suggest that the second type of synonymy corresponds to cases of somatic mutations. A similar situation has previously been reported in Cataluña for 'Col de Dame blanche', 'Col de Dame grise' and 'Col de Dame noire' which are genetically and morphologically identical and only differ by skin color [26] and in Slovenia for 'Green Matalon' and 'Black Matalon' [27]. Such mutations have been reported in *Vitis vinifera* [28], and indeed, in Brazil, a single wine producer successfully selected 2 clonal color variants [29]. In our study, each time we encountered several color forms within a variety, they occurred within the same zone, but not necessarily within the same site (see Figure 1and Additional File 3). This suggests that varieties have a prolonged local history.

We grouped several variety names as highly similar because they had the same meaning albeit in different languages or dialects (see Additional File 4). For instance, the names 'Ikoran Ihebchan -IVA1-T3P041', 'Kahla-VIA1-T4P177', 'Kohli-IA2-T8P018', and 'Taberchante-VA1-T1-P077' sampled in the central region, in the oases, in the North west and in the Moulouya valley, respectively, all corresponded to black figs presenting turbinate fruit shape, but their genotypes were distinctive. Thus cases of homonymy involved 31 distinct denominations corresponding to 181 fig trees and 147 genotypes (see Additional File 4). In a number of cases such homonymy corresponded to highly similar genotypes. Nevertheless, the denominations representing most cases of homonymy were referring to fruit color. Denominations referring to White, Black and Green color represent a total of 55 genotypes, i.e. 1/3 of the 164 genotypes sampled with variety denomination.

Depending on the genetic relationships between genotypes, three types of homonymy were distinguished (see Additional File 4). First we observed homonymy between highly similar genotypes (= within a clone) such as within the varieties 'Rhoudane', 'Zerki' and 'Byed', which included respectively three, four, and four very closely related genotypes. As stated above these correspond most probably to cases of somatic mutation within clone, and do not really constitute cases of homonymy. Second we observed cases of homonymy grouping varieties presenting similar pomological traits but clearly distinct genotypes, such as the cultivars 'Aïn Hajla', 'Rhoudane', 'Kehla' and 'Biyadi', representing respectively two, six, eight, nine and six distinct genotypes. Finally we observed cases of homonymy grouping varieties presenting different pomological traits and different genotypes (six cases; Additional File 4).

Only eight clones were present in several geographic zones. This was the case for instance for 'Assel-IA1-T4-P010' (North west zone), 'Assal-IID1-T6-P010' (Rif zone), 'Zerka-VA2-T1-P010' (Moulouya valley). These eight non local clones corresponded to widely known varieties, such as 'Assal-IID1-T6-P010' = 'Sebtawi-IA1-T1-P010' = 'Zerka-VA2-T1-P010'; 'Rhoudane-IIF1-T12-P006' = 'Rhoudani-IIIA1-T4-P006' = 'El Kehla (Rhoudani)-VC1-T1-P006' and 'Bacora-IA3-T5-P019' = 'Lemdar-IIC1-T5-P019' (see Additional File 3). Hence, in Morocco, most fig varieties are cultivated over a limited spatial. Concurrently, within a geographical zone, varieties often correspond to a single specific clone. For instance, in the Rif, the 81 trees analyzed were assigned to 43 named varieties (and 7 unnamed) and corresponded to 64 genotypes (grouped into 35 clones when including within a clones all genotypes that differed by at most three alleles).

Hence in traditional Moroccan agroecosystems fig local varieties are clones and they are generally highly local and diversified (on average 8 local varieties were collected per site in the Rif region). At least some of these local varieties were sufficiently old to have accumulated somatic mutations on neutral genetic markers and on selected traits.

2.3 Genetic diversity within and among geographical groups

Similar numbers of alleles were observed within each geographic zone, except the North center zone which presented fewer varieties, few local genotypes and as a consequence fewer alleles (Table2). Surprisingly in the South zone, all genotypes were local and allele diversity was similar to that observed in other zones. Among the 95 observed alleles, three were exclusively detected in the center zone (MFC3-133, LMFC30-259, LMFC28-192), two in the Moulouya valley (MFC9-188, LMFC24-278), two in the North west zone (LMFC19-306, LMFC32-225) and four in the South zone (MFC3-96, MFC2-190, MFC9-211, LMFC30-243). Expected heterozygosity was highest in the South zone (0.558) and lowest in the Rif zone (0.495).

There is no published data available on fig genetic diversity in traditional agroecosystems based on a sufficient number of genetic markers to discriminate clones. However, ongoing work in Lebanon and in the Tizi Ouzou area (Algeria) using the same markers (Chalak, **pers**. comm.; Daoudi, **pers**. comm.) suggest the presence of similar level of diversity as in Northern Morocco. These areas correspond to traditional agroecosystems mainly based on subsistence agriculture, with orchards presenting several fruit species grown together and several varieties per species [30, 31]. Hence, the pattern observed for fig variety diversity in Morocco can probably be transcribed to most traditional agroecosystems around the Mediterranean. How the pattern may shift outside the range of wild*Ficus carica* remains an open question.

Genetic differentiation among the six geographic zones was about 4% (F_{ST}= 0.038). Pairwise comparisons showed contrasted F_{ST}values ranging from 0.017 to 0.068 (Table 3). The highest differentiation (F_{ST}= 0.07) was noted between the Southern zone and the Rif zone. These two zones were also the sole zones clearly separated on the two first coordinate axes of the Factorial Correspondence Analysis (Figure 4). A significant spatial genetic structure was observed ($p < 10^{-6}$). Pairwise Loiselle kinship coefficients decreased significantly with distance (Figure 5), and were more strongly correlated with log than with linear distance, whatever the range of distances incorporated in the calculus. Such a pattern would be interpreted in natural populations as isolation by distance with no rupture in gene flow [32].

Table 2. Genetic diversity within geographical zone.

Geographic zone	trees analyzed	named varieties	unnamed varieties	genotypes	local genotypes	N	N_A	H_E	H_O	F_{IS}	p-value
North west (I)	60	43	4	36	31	66	3.88	0.523	0.559	-0.0699	0.0187
Rif (II)	81	43	7	64	58	70	4.12	0.495	0.540	-0.0968	0.0003
Mountain agroecosystems [a]	141	76	11	96	89	77	4.53	0.510	0.548	-0.0724	0.0005
North center (III)	20	12	4	15	12	54	3.18	0.533	0.558	-0.0502	0.1477
Center (IV)	61	23	10	45	40	70	4.12	0.511	0.557	-0.0904	0.0156
Moulouya valley (V)	34	19	2	27	24	70	4.12	0.518	0.507	0.0219	0.4292
South (VI)	21	15	3	21	21	72	4.24	0.558	0.571	-0.0242	0.2898
Oasis agroecosystems [b]	21	14	3	21	21	72	4.24	0.558	0.571	-0.0242	0.2898

N: total number of alleles observed within each zone; N_A: mean number of alleles per locus; H_O: observed heterozygosity; H_E: expected heterozygosity; F_{IS}: intrapopulation fixation index. [a] Moutain agroecosystems (= Rif and North west zones); [b] Oasis agroecosystems (= oases of the South Morocco).

Table 3. Pairwise F_{ST} values between samples from the different geographic zones.

	North west	Rif	North center	Center	Moulouya
Rif	0.028**				
North center	0.026*	0.046**			
Center	0.021***	0.030***	0.025*		
Moulouya	0.027***	0.031**	0.026*	0.017**	
South	0.038***	0.068***	0.042*	0.018*	0.029**

* $p < 5.10^{-3}$, ** $p < 10^{-6}$, *** $p < 10^{-9}$

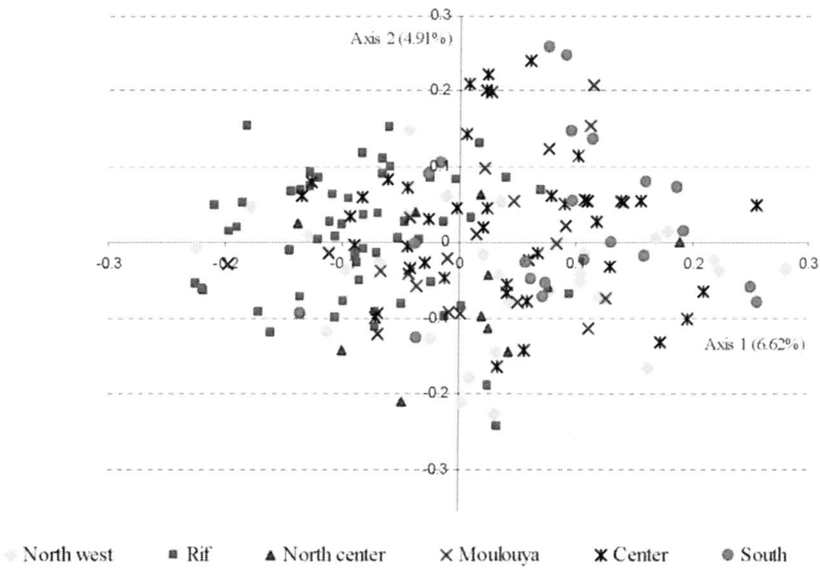

Figure 4. Separation of genotypes according to zone of origin on the two first axes of the Factorial Correspondence Analysis. The Southern zone (in red) and the Rif zone are separated (in blue).

We may reconcile the three sets of analyses (FCA, F_{ST} and pairwise Loiselle kinship coefficients) by suggesting that we have here the image of spatial genetic structure as could be expected in natural populations for a situation of mutation/migration/drift processes at equilibrium resulting in some geographic variation in genetic background without geographic variation in genetic diversity. Within this global pattern, the North west zone appears to be slightly atypical, a feature which could have been predicted. Indeed, the region is the most affected by neighboring cities and as such represents a less traditional agroecosystem, slightly blurring the picture.

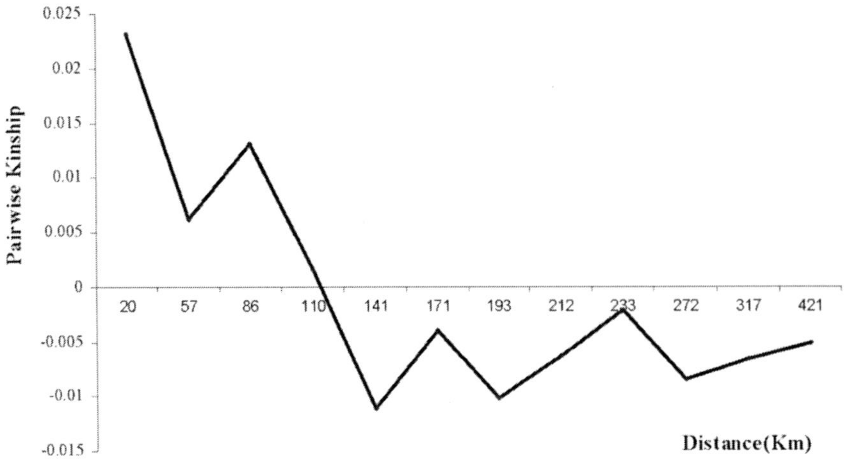

Figure 5. Pairwise kinship coefficient between genotypes as a function of geographic distance.

The pattern of isolation by distance, with no clines in diversity, is a signature of a genetic equilibrium situation, with no trace of a past colonization process. This feature and the quasi-absence of widespread varieties, is suggestive of a cultivation system based on varieties that originated locally, mainly from the local gene pool.

2.4 Varietal and genotypic diversity in mountain and oasis agroecosystems

Quite interestingly, traditional mountain agroecosystems (North west and Rif zones) presented much more varietal diversity than traditional oasis agroecosystems (South zone) (Table 2, Figure 2Band 2C). However they presented almost identical numbers of alleles. This result was obtained despite our sampling only 21 trees in the oases against 141 trees in the North west and Rif zone. This suggests that fig varietal and genetic diversity available in oases is threatened, maybe due to their small surface, while the one available in the mountain agroecosystems will be more resilient.

3. CONCLUSIONS

Traditional Moroccan agroecosystems contain substantial fig varietal and genetic diversity. While fig varieties are true clones and not landraces [16], the distribution of differences between genotypes shows that this diversity arose through sexual reproduction and only marginally, through somatic mutation. Hence the silver bullet hypothesis of instantaneous domestication of clonal plants [13] does not apply, at least today, to fig. In that perspective fig is similar to other clonally propagated plants from other parts of the world for which sexual reproduction has been important and often still is. Such species include for instance Cassava [33] and Agave [34] in America or Enset [35] in Africa. Further, in fig, sexual production of new varieties almost obligatorily involves crosses with wild figs. Indeed, it is a dioecious species, and male figs used for pollination are collected on any tree in the neighborhood, and when male figs are cultivated within a village, their potential genetic qualities for siring agronomically interesting crops is not taken into account. Preliminary data from the Rif zone confirms close genetic relationship between local varieties and wild growing fig trees. As such fig cultivation in its native range fits the global picture of frequent hybridization of cultivated plants with their wild relative [36].

However the case of fig is particular as new varieties must (almost) systematically result in the incorporation of hybrids between wild and cultivated plants. We may thus suspect that in all traditional Mediterranean agroecosystems located within fig natural habitat, cultivated figs and wild growing figs locally form a single evolutionary unit. Hence such traditional agroecosystems are effectively incubators of fig variety diversity in a dynamic incorporating wild growing as well as cultivated trees. This is not always the case in clonally propagated plants. For instance, while sexual reproduction seems to be most important in traditional Cassava cultivation, genetics allow to trace its origin to a single region of the range of its progenitor, *Manihot flabellatus* [37]. The domestication process of monoecious and dioecious plants may turn out to be quite different.

In a context of ongoing rapid climatic change, the nutritional quality, and toxicity of crops may change dramatically [38]. A dynamic management of genetic resources as observed here in traditional agroecosystems may prove essential for responding to such new challenges.

4. METHODS

4.1 Fig sampling

Traditional agroecosystems are still present in Morocco, in the Rif and Atlas mountains in Northern and central areas and in oases in the South east. A survey in the Rif agroecosystems showed that 28 crop species were cultivated including 14 fruit species [31]. A high diversity of fruit crops was also observed in the South Moroccan oases.

Field trips to collect plant material covered all territories of Morocco presenting traditional agroecosystems (Figure 1). They were done in June and August-September in order to observe first or second crop figs, respectively (fig varieties produce either both first crop and second crop or only the second crop). Collections were made in 2005 and 2006. This allowed characterizing the different varieties and establishing their geographical range. Field observations and some genetic data (Achtak et *al.* unpublished) had shown that within the range of each prospection site or village, each variety corresponded generally to a single genetic clone. The sampling strategy could therefore be focused on diversity, using pomological observation following the IPGRI recommendation [39] and interviews with farmers. Thus, for each prospection site, we sampled one individual of each of the cultivated varieties. When we had a doubt on the perfect identity of vegetative and pomological traits within a variety within a site, or when a farmer suggested that there were two types within a variety, then we collected both forms. Hence genetic homogeneity within variety was assessed within site when there was any hint of a doubt, and systematically, among sites. Local variety names were noted as given by farmers; photographs and GPS coordinates were recorded as references for each collected fig tree (see Additional File 1). The photographs allowed confronting a posteriori genotypic identity with morphological similitude. Six major geographical zones were surveyed (North west, Rif, North center, Center, Moulouya valley and South; Figure 1) and 277 trees representing 119 denominations were sampled.

4.2 DNA extraction and SSR genotyping

Total genomic DNA was extracted from 200 mg of fresh young leaves of the 277 sampled fig trees using the DNeasy Plant Mini Kit (QIAGEN) according to the supplier's instructions with the following modification: 1% of Polyvinylpyrrolidone (PVP 40,000) was added to the buffer AP1.

We selected 17 loci among the developed SSR markers [22–25] based on their polymorphism and ease of scoring following the screening of 16 distinct Mediterranean varieties.

Microsatellite amplifications were performed according to the protocol described by Khadari et *al.* [40]. SSR genotyping was conducted in an automated capillary sequencer (ABI prism 3130 XL). Analyses were performed using the GENEMAPPER V3.7 software.

4.3 Data analysis

For each SSR locus, alleles were detected and identified by locus name and allele size in bp. Genetic distances between fig genotypes were estimated according to the Jaccard similarity coefficient and UPGMA algorithm using a program developed by J. Brzustowski http://www.biology.ualberta.ca/jbrzusto/cluster.php. The corresponding phenogram was drawn based on the software Treeview 6.1. Discriminating power, D, was calculated for each SSR locus as $D_j = \sum p_i [(Np_i\text{-}1)/N\text{-}1)]$ [41] where p_i was the frequency of the *i*-th molecular pattern revealed by locus *j*, and *N* was the number of genotypes. We used the *Dj* values to compute the exact probabilities of getting at least one pair of genotypes differing only at 0, 1, 2, 3, 4, 5 and 6 loci.

The number of alleles per locus (*A*), observed heterozygosity (H_O), expected heterozygosity (H_E) and Wright's fixation index ($F = 1\text{-}H_O/H_E$) were computed using the software Genetix 4.5 [42]. Genetic diversity was compared among geographic zones using parameters corrected for sample size [43]. Genetic differentiation between populations was assessed using F_{ST} values and the software Genepop 3.1 [44]. The significance of population differentiation was estimated using exact tests [45].

To assess genetic isolation by distance, spatial genetic structure was investigated using a spatial autocorrelation method. Genetic relationships between all pairs of genotypes were regressed on the linear and the logarithmic geographical distance using the software SPAGeDi [46]. The kinship coefficient of Loiselle *et al.* [47], robust against the presence of low frequency alleles, was used. Significance of the regression coefficients was assessed through 10,000 permutations.

AUTHORS' INFORMATION

HA is a geneticist who defended the PhD thesis entitled "domestication and diversification of fig,*Ficus carica* L., varieties in Morocco" in December 2009. MA is a Professor at University of Tétouan (Morocco), plant ecologist and working on agrobiodiversity. AO is a scientist at the Moroccan Agronomic Research Institute in charge of the management of fruit genetic resources. SS is an engineer, molecular biologist, in charge of the development and use of molecular markers. FK is a scientist in evolutionary biology particularly on biotic interactions, specialised on the *Ficus*pollinating wasps co-evolution. BK is a geneticist working within the conservation biology field and managing a Mediterranean fruit domestication and diversification program.

ACKNOWLEDGEMENTS

The PhD student, Hafid Achtak, was supported by fellowships from Agropolis Fondation, RTRA N° 07042 "FigOlivDiv", PRAD 04-07, GDRI BIOM and Volubilis Ma-08-197. This work was performed in the laboratory "Atelier de Marquage Moléculaire" UMR DiA-PC, INRA Montpellier and was financed by GDRI BIOM and Agropolis Fondation, RTRA N° 07042 "FigOlivDiv". We thank all members of the laboratory "Atelier de Marquage Moléculaire" for their technical support. During the whole study, the encouragement and help of Pr. F. Dosba and Dr. F. Boillot were decisive. Comments on an earlier version of the manuscript from Pr. Doyle Mckey, Loic Le Cunff and Jean-Louis Noyer seriously improved its final quality.

AUTHORS' CONTRIBUTIONS

BK designed and coordinated the study. HA, MA, AO and BK performed fig sampling. HA, SS and BK carried out the molecular analysis. HA, FK and BK performed the statistical analysis and wrote the first draft of the manuscript. All authors participated in the draft finalization and approved the final manuscript.

REFERENCES

1. Esquinas-Alcazar J: Protecting crop genetic diversity for food security: political, ethical and technical challenges. Nature Reviews Genetics. 2005, 6: 946-953.

2. Mooney H, Cropper A, Reid W: Confronting the human dilemma: How can ecosystems provide sustainable services to benefit society?. Nature. 2005, 434: 561-562.

3. Myers N, Mittermeier RA, Mittermeier CG, Da Fonseca GAB, Kent J: Biodiversity hotspots for conservation priorities. Nature. 2000, 403: 853-858.

4. Brush SB: In situ conservation of landraces in centers of crop diversity. Crop Science. 1995, 35: 346-354.

5. Zhu Y, Chen H, Fan J, Wang Y, Li Y, Chen J, Fan J, Yang S, Hu L, Leung H, Mew TW, Teng PS, Wang Z, Mundt CC: Genetic diversity and disease control in rice. Nature. 2000, 406: 718-722.

6. Gebauer J, Luedeling E, Hammer K, Nagieb M, Buerkert A: Mountain oases in northern Oman: An environment for evolution and in situ conservation of plant genetic resources. Genetic resources and crop evolution. 2007, 54: 465-481.

7. Jarvis D, Brown AHD, Cuong PH, Panduro CL, Moreno LL, Gyawali S, Tanto T, Sawadogo M, Mar I, Sadiki M, Hue NTN, Reyes AL, Balma D, Bajracharya J, Castillo F, Rijal D, Belqadi L, Rana R, Seddik S, Ouedraogo J, Zangre R, Rhrib K, Chavez JL, Schoen D, Sthapit B, De Santis P, Fadda C, Hodgking T: A global perspective of the richness and evenness of

traditional crop-variety diversity maintained by farming communities. PNAS. 2008, 105: 5326-5331.

8. Hillebrand H: On the generality of the latitudinal diversity gradient. American Naturalist. 2004, 163: 192-211.

9. Mittelbach GG, Schemske DW, Cornell HV, Allen AP, Brown JM, Bush MB, Harrison SP, Hurlbert AH, Knowlton N, Lessios HA, McCain CM, McCune AR, McDade LA, McPeek MA, Near TJ, Price TD, Ricklefs RE, Roy K, Sax DF, Schluter D, Sobel JM, Turelli M: Evolution and the latitudinal diversity gradient: speciation, extinction and biogeography. Ecology Letters. 2007, 10: 315-331.

10. Médail F, Quézel P: Biodiversity hotspots in the Mediterranean basin: Setting global conservation priorities. Conservation Biology. 1999, 13: 1510-1513.

11. Willcox G, Buxo R, Herveux L: Late Pleistocene and early Holocene climate and the beginnings of cultivation in northern Syria. Holocene. 2009, 19: 151-158.

12. Weiss E, Kislev ME, Hartmann A: Autonomous cultivation before domestication. Science. 2006, 312: 1608-1610.

13. Zohary D: Unconscious selection and the evolution of domesticated plants. Economic Botany. 2004, 58: 5-10.

14. Kislev ME, Hartmann A, Bar-Yosef O: Response to comment on "Early domesticated fig in the Jordan valley". Science. 2006, 314: 1683-

15. Lev-Yadun S, Ne'eman G, Abbo S, Flaishman MA: Comment on "Early domesticated fig in the Jordan valley". Science. 2006, 314

16. Harlan JR: *Crops and Man*. American Society of Agronomy, Madison, WI. 1, 1975.

17. Kislev ME, Hartmann A, Bar-Yosef O: Early domesticated fig in the Jordan Valley. Science. 2006, 312: 1372-1374.

18. Kjellberg F, Gouyon PH, Ibrahim M, Raymond M, Valdeyron G: The stability of the symbiosis between dioecious figs and their pollinator: a study of *Ficus carica* L. and *Blastophaga psenes* L. Evolution. 1987, 41: 693-704.

19. Condit IJ: Fig varieties: a monograph. Hilgardia, Berkeley. 1955, 23: 323-538.

20. Aumeeruddy-Thomas Y, Hmimsa Y, Ater M, Khadari B: Beyond the divide between wild and domestic: domesticity, spatiality and practices pertaining to fig (*Ficus carica*) and olive (*Olea europaea*) agroecosystems in Morocco. Crops and people: choices and diversity through time. Edited by: Chevalier A, et al. Oxfam, London,

21. Khadari B, Grout C, Santoni S, Kjellberg F: Contrasting genetic diversity and differentiation among Mediterranean populations of *Ficus carica* L.: a study using mtDNA RFLP. Genetic resources and crop evolution. 2005, 52: 97-109.

22. Khadari B, Hochu I, Santoni S, Kjellberg F: Identification and characterisation of microsatellite loci in the common fig (*Ficus carica* L.) and representative species of genus *Ficus*. Molecular Ecology Notes. 2001, 1: 191-193.

23. Achtak H, Oukabli A, Ater M, Santoni S, Kjellberg F, Khadari B: Microsatellite markers as reliable tools for fig cultivar identification. Journal of American Society of Horticultural Science. 2009, 134: 624-631.

24. Ahmed S, Dawson DA, Compton SG, Gilmartin PM: Characterization of microsatellite loci in the African fig*Ficus sycomorus* L. (*Moraceae*). Molecular Ecology Notes. 2007, 7: 1175-1177.

25. Giraldo E, Viruel MA, López-Corrales M, Hormaza JI: Characterization and cross-species transferability of microsatellites in the common fig (*Ficus carica* L.). Journal of Horticultural Science and Biotechnology. 2005, 80: 217-224.

26. Khadari B, Lashermes P, Kjellberg F: RAPD fingerprints for identification and genetic characterization of fig (*Ficus carica* L.) genotypes. Journal of Genetics and Breeding. 1995, 49: 77-86.

27. Bandelj D, Jakse J, Javornik B: Development of microsatellite markers for identification of fig varieties in Istria. The common fig (Ficus carica L.) in Istria (morphological, molecular and some chemical characteristics)Koper, Slovenia: Annales Publishing HouseBandelj D, Miklavcic MB, Vrhovnik I. 2008,84-89.

28. Yakushiji H, Kobayashi S, Goto-Yamamoto N, Tae Jeong S, Sueta T, Mitani N, Azuma A: A skin color mutation of grapevine, from black-skinned Pinot Noir to white-skinned Pinot Blanc, Is caused by deletion of the functional *VvmybA1* allele. Bioscience, Biotechnology, and Biochemistry. 2006, 70: 1506-1508.

29. De Oliveira-Collet SA, Zequim-Maia SH, Mangolin CA, de Fátima Pires da Silva Machado M: Mutações e Recombinações Genéticas Pesquisa geram uvas coloridas. Biotecnologia Ciência & Desenvolvimento. 2005, 35: 28-35.

30. Taiqui L, Cantarino CM: Elément historiques d'analyse écologique des paysages montagneux du Rif Occidental (Maroc). Mediterranea Serie de estudios biologicos. 1997, 16: 23-36.

31. Hmimsa Y, Ater M: Agrodiversity in the traditional agrosystems of the Rif mountains (North of Morocco). Biodiversity. 2008, 9: 78-81.

32. Vekemans X, Hardy OJ: New insights from fine-scale spatial genetic structure analyses in plant populations. Molecular Ecology. 2004, 13: 921-935.

33. Rival L, McKey D: Domestication and diversity in Manioc (*Manihot esculenta* Crantz ssp. *esculenta*, Euphorbiaceae). Current Anthropology. 2008, 49: 1119-1128.

34. Parker KC, Hamrick JL, Hodgson WC, Trapnell DW, Parker AJ, Kuzoff RK: Genetic consequences of pre-columbian cultivation for *Agave murpheyi* and *A. delamateri* (Agavaceae). American Journal of Botany. 2007, 94: 1479-1490.

35. Birmeta G, Nybom H, Bekele E: RAPD analysis of genetic diversity among clones of the Ethiopian crop plant*Ensete ventricosum*. Euphytica. 2002, 124: 315-325.

36. Jarvis DI, Hodgkin T: Wild relatives and crop cultivars: detecting natural introgression and farmer selection of new genetic combinations in agroecosystems. Molecular Ecology. 1999, 8: S159-S173.

37. Leotard G, Duputié A, Kjellberg F, Douzery EJP, Debain C, De Granville JJ, Mckey D: Phylogeography and the origin of cassava: new insights from the northern rim of the Amazonian basin. Molecular Phylogenetics and Evolution. 2009, 53: 329-334.

38. Gleadow RM, Edwards EJ, Evans JR: Changes in nutritional value of cyanogenic Trifolium repens grown at elevated atmospheric CO_2. Journal of Chemical Ecology. 2009, 35: 476-478.

39. IPGRI, CIHEAM: Descriptors for fig. , International Plant Genetic Resources Institute, Rome, Italy, and International Centre for Advanced Mediterranean Agronomic Studies, Paris, France 2003, 52.

40. Khadari B, Oukabli A, Ater M, Mamouni A, Roger JP, Kjellberg F: Molecular characterization of Moroccan fig germplasm using Intersimple Sequence Repeat and Simple Sequence Repeat markers to establish a reference collection. HortScience. 2004, 40: 29-32.

41. Tessier C, David J, This P, Boursiquot JM, Charrier A: Optimization of the choice of molecular markers for varietal identification in *Vitis vinifera* L. Theoretical and Applied Genetics. 1999, 98: 171-177.

42. Belkhir K, Borsa P, Chikhi L, Raufaste N, Bonhomme F: GENETIX 4.05, logiciel sous Windows TM pour la génétique des populations. Laboratoire Génome, Populations, Interactions, CNRS UMR 5000, Université de Montpellier II, Montpellier, France. 1996

43. Nei M: Estimation of average heterozygosity and genetic distance from a small number of individuals. Genetics. 1978, 89: 583-590.

44. Raymond M, Rousset F: GENEPOP (version 1.2): population genetics software for exact tests and ecumenicism. Journal of Heredity. 1995, 86: 248-249.

45. Raymond M, Rousset F: An exact test for population differentiation. Evolution. 1995, 49: 1280-1283.

46. Hardy O, Vekemans X: SPAGeDi: a versatile computer program to analyse spatial genetic structure at the individual or population. Molecular Ecology Notes. 2002, 2: 618-620.

47. Loiselle BA, Sork VL, Nason J, Graham C: Spatial genetic structure of a tropical understory shrub, *Psychotria officinalis* (Rubiaceae). American Journal of Botany. 1995, 82: 1420-1425..

Chapter 10

Variation in Catechin Composition of Popularly Cultivated Tea Clones in East Africa (Kenya)

Leonida Cherotich[1], Samson M. Kamunya[2], Amos Alakonya[1], Solomon W. Msomba[3], Marie A. Uwimana[4], John K. Wanyoko[2], Phillip O. Owuor[5]

[1]Institute of Biotechnology Research (IBR), Jomo Kenyatta University of Agriculture and Technology, Nairobi, Kenya

[2]Tea Research Foundation of Kenya, Kericho, Kenya

[3]Tea Research Institute of Tanzania, Dar Es Salaam, Tanzania

[4]Office Des Cultures Industrielles du Rwanda The' (OCIR The'), Kigali, Rwanda

[5]Department of Chemistry, Maseno University, Maseno, Kenya.

ABSTRACT

Tea polyphenols and catechins have gained prominence in determination of tea quality. Tea leaves contain large amounts of catechins and their derivatives and these have shown positive effects on human health. An experiment was carried out to determine the level of catechins and polyphenols in 20 new clones found in two distinct tea growing areas. The trial comprising twenty clones was established in two distinct sites in Kenya using randomized complete block design with three replicates and consisted of 10 popular cultivars widely cultivated in the Lake Victoria Basin and additional 10 newly improved clones. Tea leaf samples were collected across two seasons; dry and wet seasons. Tea quality parameters entailing catechins, caffeine and total polyphenols was generated for the 20 clones across the two seasons and sites. The ISO 14502-1 and ISO 14502-1:2005 procedures were used for polyphenol and catechin analyses, respectively. The ANOVA results generated indicated that the level of catechins and polyphenols in tea were significantly different ($p < 0.05$) and varietal dependent. Generally, the assam varieties had higher levels of catechins and polyphenols than the chinary and cambod varieties, indicating why they are popularly cultivated in the East African region and whose tea quality is highly regarded. Furthermore, the results from principle component analysis (PCA) grouped the clones into three major clusters with one comprising of high quality assam varieties, a mixture of moderate quality assam, cambod and chinary varieties and low catechin content assam anthocyanin-rich variety. It is concluded that the level of catechins in tea plants are varietal dependent, although environmental stress has effects on the synthesis of flavanoids. The results demonstrated that the differences in the accumulation of catechins among the tea clones could be important for the discrimination of clonal diversity.

KEYWORDS

Tea; Health; Camellia sinensis; Catechins; Variation; Kenya

1. INTRODUCTION

Tea, camellia sinensis (L) O. Kuntze, the most popular universal beverage contains a wide range of the Flavanoids [1]. The main flavanoid compounds found in tea are: epigallocatechin (EGC), epicatechin (EC), epigallocatechin gallate (EGCG), epicatechin gallate (ECG) and other compounds [2]. Flavonoids in tea have been reported to be the major source of antioxidants, which have a variety of physiological functions to the body [3]. Catechins in tea have been reported to promote energy consumption by activating Beta-oxidative enzymes of lipids and so weight loss [4]. Reportedly, catechins in tea, especially (-)-epigallocatechin gallate (EGCG), inhibit the angiotensin converting enzyme, suppress the production of angiotensin II, which has strong hypertensive activity, and induce antihypertensive activity [5]. Catechin compounds have been of focus for an anticancer [6]. Biochemicals in tea also relieve allergy symptoms and are shown to improve the body's immunity [7]. Studies have shown that catechins in tea inhibit diabetes including hyperglycaemia by reducing elevated sorbitol decreasing protein glycosylation and lipid peroxidation and also through inhibiting diabetic cataracts [8]. Catechin compounds have been proven to have a variety of physiological functions, such as those enacting in the duodenum, colon, skin, lung, breast, oesophageal, pancreatic and prostate cancer function [6]. In addition to prophylactic activity, catechins have been found to inhibit the growth of food borne pathogenic bacteria, and do not have adverse effects on the beneficial bacteria [3].

Given the importance of tea to the economy of many countries and human health, the use of suitable tea cultivars that are well endowed in both yield and phytochemicals like flavonoids is paramount. The tea improvement programme at the Tea Research Foundation of Kenya (TRFK) has resulted in many new tea varieties, but exploitation of only a few popular clones accounting for about 60% of tea in the country may have resulted in narrowing of the clones which could be exploited for the commercializable germplasm [9]. Although the level of genetic relatedness of these clones has been elucidated by DNA markers, the application of their biochemical interrelationships remains largely unknown. As has been demonstrated by previous studies [10], biochemical composition of tea plant can be used as markers to elucidate diversity of the tea clones. The green tea leaf contains 30 to 42% polyphenols on the dry weight basis while a cup of green tea contains about 30% to 40% of polyphenols [11]. Polyphenols have aromatic rings with multiple hydroxyl residues. The major polyphenols in tea are the catechins [12]. This group of compounds is derived from the phenylpropanoid and flavanoid biosynthetic pathways. Different tea clones have different number, relative amounts and diversity in their level of catechins. An understanding of the biochemical composition in the plant system might yield information for plant genetic manipulation and crop management strategies that might improve crop value in future [13]. Catechins also play a key role in quality determination in black tea and fruits [14]. The aim of this study was to determine the level of catechins in the 20 improved clones of tea, 10 of which are widely cultivated in the East African tea growing areas.

2. Methodology

2.1. Description of Study Sites

The trial comprising twenty clones was established in April 2009 in two major tea growing areas: at the Timbilil Estate of the TRFK, Kericho and Kipkebe in Sotik Ltd (**Table 1**). The trial was set up in a randomized complete block design with three replications in plots of 30 plants spaced at 0.61 m within rows and 1.22 m between rows (i.e. 13,448 plants/ha). The trial received 150 Kg N per hectare per year in the form of NPK 25:5:5 compound fertilizer. Each replicate was surrounded by a guard row of clone TRFK 301/4. The tea was brought into bearing following the recommended agronomic practices [7].

2.2. Description of Genetic Materials

The trial consisting of 10 popular cultivars widely cultivated in the Lake Victoria Basin covering Kenya, Rwanda, Uganda and Tanzania and an additional 10 newly improved clones was initiated in 2009 as one of two separate studies entailing "Evaluation of tea cultivars to different growing environments in East African countries" and "Evaluation of response of clone TRFK 6/8 to nitrogen rates and harvesting intervals in East African countries" [15]. The current study however, covered the two Kenyan sites. The cuttings from the mother bushes were propagated in April 2008 and reared in the nursery for one year. Nursery materials were transplanted into the field in April/May 2009 at Kipkebe (Sotik) and Timbilil (Kericho) and brought into bearing following the recommended agronomic practices [7]. The various attributes and status of the clones are summarized in the**Table 2**.

2.3. Laboratory Analysis of Catechins and Polyphenols

2.3.1. Leaf Sampling and Sample Processing

About 500 g of fresh shoots in form of two leaves and a bud were plucked from each of the clonal entries and placed in appropriately labelled khaki bags. These were then transferred into a cool box containing ice packs and transported to the lab for processing. The clones were sampled twice under different environmental conditions (wet and dry seasons) as shown in **Table 3**. The samples were transported to the laboratory and immediately dried using a microwave for 5 minutes, which also deactivated the oxidizing enzyme polyphenol oxidase. The dry samples were then ground using a coffee miller and placed in aluminium bags, before moving them to a dark, dry environment for later laboratory analysis [10].

2.3.2. Extraction and Purification of Catechins and Polyphenols

Catechin analysis was done following the protocol and the procedure described by the International Organization for Standardization (ISO) 14502-1:2005. Two hundred milligrams of ground samples were extracted with 5 ml of warm (70˚C) 70% methanol. The mixture was then warmed in a water bath (70˚C) for 10 minutes and vortexed in between after five minutes interval and then cooled. The cool mixture was centrifuged at 3500 rpm for 10 minutes, and the extract was decanted into a 10 ml cylinder. The samples were then reextracted and the extract decanted to the initial 10 ml cylinder topped-up to 10 ml. One millilitre of the extract was diluted 5-fold using stabilizer reagent and passed through a 0.5 mm pore filter before injection into the HPLC C-6 column for catechin analysis. The mobile phase A consisted of 9% (volume fraction) acetic acid, and mobile phase B was made of 80% (volume fraction) acetonitrile. The flow rate was controlled at 1.0 ml/min, with an injection volume of 20 ml. The column was operated at 40˚C and UV spectra peaks were detected at 278 nm. The chromatographic peaks in the samples were identified by comparing their retention times with chemical standards used.

Table 1. Location, elevation and climatic characteristics of the two experimental sites in Kenya.

Site	Latitude	Longitude	Elevation (m)	Annual average temperature (˚C)	Annual rainfall (mm)	Solar radiation ($MJ \cdot m^{-2} \cdot d^{-1}$)
Timbilil	0˚22'S	35˚21'E	2180	16.28	2043	20.72
Kipkebe	0˚41'S	35˚5'E	1800	20.1	2,100	-

Table 2. Attributes of the 20 clones under investigation found in two distinct areas.

S.N.	Clone	Varietal type	Special attributes	Status
1	TRFK 12/19	Assam type local selection	High black tea quality, moderate yields	Widely grown in EA
2	TRFK 301/4	Cambod type local selection	Acceptable black tea quality, drought tolerant and high yielding	Widely grown in Kenya and recently introduced in Rwanda and Tanzania
3	EPK C12	Chinary type local selection	Acceptable black tea quality, drought tolerant and moderate yielder	Widely grown in Kenya
4	TRFK 301/6	Cambod type local selection	Drought tolerant and moderate yielder	Test clone yet to be released for commercial use
5	TRFK 31/8	Assam type local selection	Acceptable black tea quality, high yielding	Widely grown in EA
6	TRFK 31/11	Assam type local selection	High black tea quality, moderate yielder	Widely grown in EA
7	TRFK 12/12	Assam type local selection	High black tea quality, moderate yielder	Widely grown in EA
8	TRFK 306	Assam type purple tea	Moderate yielder and medicinal properties	Released in 2011 for specialty tea product
9	BBK 35	Assam type local selection	High black tea quality, moderate yielder	Widely grown in EA
10	TRFK 430/90	Assam type local hybrid	High black tea quality, high yielding	Recently released in Kenya for commercial use
11	TRFK 6/8	Assam type local selection	High black tea quality, moderate yielder	Widely grown in EA
12	TRFK 303/577	Assam/China hybrid, OP of 6/8	High black tea quality, high yielding	Widely grown in EA
13	TRFK 303/1199	Assam/China hybrid, OP of 6/8	High black tea quality, high yielding	Widely grown in EA
14	TRFK 301/5	Cambod type local selection	Acceptable black tea quality, root knot nematode tolerant and high yielding	Widely grown in Kenya and recently introduced in Rwanda and Tanzania
15	TRFK 7/3	Assam type local selection	High black tea quality, moderate yielder	Widely grown in EA
16	TRFK 6/10	Assam type local selection	Acceptable black tea quality, moderate yielder	Widely grown in EA
17	TRFK 371/3	Assam type local hybrid	High black tea quality, high yielding	Recently released in Kenya for commercial use
18	TRFK 100/5	Assam type local selection	High black tea quality, moderate yielder	Widely grown in EA
19	TRFK 11/4	Assam type local selection	Acceptable black tea quality, high yielding	Widely grown in EA
20	TRFK 12/56	Assam type local selection	Acceptable black tea quality, moderate yielder	Widely grown in EA but not commercialised in Kenya

Table 3. Monthly rainfall and temperature data for Kipkebe and Timbilil during the wet and dry season.

			Wet season			Dry season		
	Site		Jul-11	Aug-11	mean	Jan-12	Feb-12	mean
Monthly total rainfall(mm)	Timbilil		165.5	169.8	167.7	0.0	26.8	13.4
	Kipkebe		100.8	273.1	187.0	0.5	82.9	41.7
Temperature (°C)	Timbilil	Max.	23.0	8.3	15.7	25.7	7.7	16.7
		Min.	22.7	9.4	16.1	26.3	9.1	17.7
	Kipkebe	Max.	26.7	26.2	26.5	28.7	28.4	28.6
		Min.	12.4	12.4	12.4	11.0	11.7	11.4

For polyphenols analysis, the method described by the International Organization for Standardization (ISO) 14502-1 was used for polyphenols. 1.0 ml of the garlic acid standard solutions were transferred into separate glass tubes labelled A, B, C, D and E in triplicates. Subsequently, 1.0 ml of the diluted sample extract was transferred into separate triplicate tubes. 5.0 ml of 10% Folin-Ciocalteu's reagent was added. Then immediately, 4.0 ml of a sodium carbonate solution (7.5% w/v) was added. The tubes were vortexed and then allowed to stand at room temperature for 60 min before absorbance was determined at 765 nm against the blanks using a spectrophotometer.

2.4. STATISTICAL ANALYSIS

The various catechins and polyphenols contents in the clones were analyzed using M-Stat-C statistical software for ANOVA. Principle component analysis (PCA) on the concentration of catechins in different clones was done using GENSTAT software [16].

3. RESULTS

3.1. Variations in Catechins among the Cultivars

The catechins were eluted in the following order Gallic Acid (GA), Epigallo catechin (EGC), Caffeine (Caff), Epicatechin (EC), Epigallocatechingalate (EGCG), Catechin(C) and Epicatechin galate (ECG). An example of eluted catechins following HPLC analysis is presented in **Figure 1**.

A recapitulation of mean squares of the measured parameters is shown in **Table 4**. It is apparent from the Table that the parameters measured showed varying levels of significance with variations attributed to clones, interactions between (clones × season), and (clone × season × site), being highly significant ($p < 0.001$). The effect due to seasons was significant ($p < 0.01$) for all but one parameter EGC. Similarly the effect due to site was significant ($p < 0.05$) for nearly all the parameters except EGCG, ECG, and TCT. The significance variation of season × clone interactions indicated that clones responded differently to different seasons and sites. The variations in levels of various catechins and total polyphenols (TPs) with respect to seasons and sites are presented in Tables 5-13.

3.1.1. Seasonal Variations of Total Polyphenols (TP) at Timbilil and Kipkebe

The TP contents in the 20 clones presented in **Table 5**, varied significantly across different sites and seasons ($p < 0.05$). The mean of the TP across four seasons also varied significantly. Considering the overall means, clone

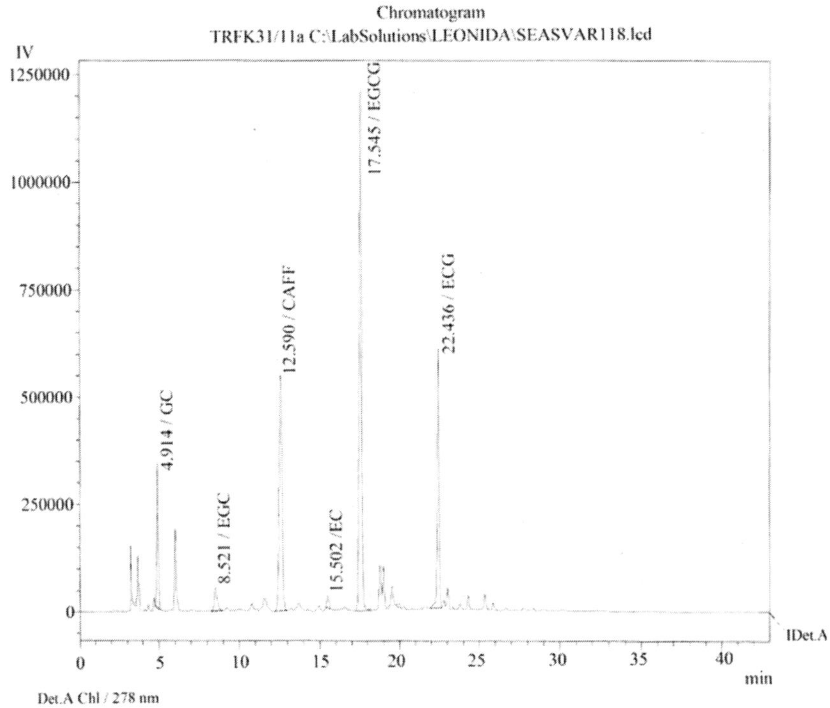

Figure 1. Chromatogram of clone TRFK 31/11 indicating the elution time and the order of the elution of the individual catechins.

TRFK 12/12, had significantly higher levels of TP at 24.52%, while clone TRFK 301/6 had the least levels, with a mean of 20.14%. However, clones TRFK 31/11, TRFK 371/ 3, BBK 35, had similar TP content at 24.02%, 23.92%, and 23.80% respectively. There was no notable differences in the mean total polyphenols across four environments for clones TRFK 301/4, TRFK 430/90, and TRFK 303/1199.

Table 4. Summary of mean squares for assessed catechins and total polyphenols.

S.V	Df	Mean squares (MS)								
		GA	EGC	C	CAFF	EC	EGCG	ECG	TCT	TP
Season	1	0.307***	0.747	0.152***	0.854***	1.697***	33.705***	1.547**	51.699***	18.984***
Site (season)	2	0.113***	2.166***	0.085***	3.532***	0.360*	1.233	0.228	0.679	81.891***
Rep (seasonxsite)	8	0.032***	0.356	0.014	0.228**	0.21	0.916	0.313*	2.905*	1.129
Clone	19	0.329***	15.829***	0.076***	3.428***	7.580***	29.274***	9.224***	41.392***	18.622***
Seasonxclone	19	0.035***	1.073***	0.031***	0.202**	1.422***	5.749***	0.293**	4.330***	3.722***
Sitexclone (season)	38	0.017***	0.688***	0.038***	0.295***	0.512***	1.480***	0.330***	3.364***	6.216***
Error	152	0.007***	0.25	0.009	0.082	0.118	0.596	0.14	1.208	1.159

NB: Asterisks *, **, and *** denote significance level at $p < 0.05$, $p < 0.01$ and $p < 0.001$ respectively.

At the Timbilil site, clone TRFK 371/3 had the highest level of TP at 26.49% during wet season, while clone EPK C12 had the least content of polyphenols at 19.81%. In the dry season, clone TRFK 303/1199 had the highest level of polyphenols of 23.84%, while clone TRFK 301/6 had the least at 18.32%. Although most of the clones showed some slight drop in total polyphenols from wet to dry seasons, the drop by clone TRFK 6/8 and TRFK 371/3 of 6% was rather drammatic.

At the Kipkebe site, clone TRKF 12/12 had the highest level of TP at 25.17% during the wet season, while clone TRFK 301/6 had the least at 19.81%. Clone TRFK 31/11 had the highest level polyphenols of 26.89%, while clone TRFK 301/6 had the least at 20.94% during dry season. Interestingly, clone TRFK 306 had the least level of TP during both seasons. It is worth noting that, unlike for Timbilil, the level of total polyphenols in Kipkebe rose on average from wet to dry season. Clones TRFK 100/5, TRFK 31/11 and TRFK 371/3 appeared to have responded more positively by the onset of stress visited by the dry season.

3.1.2. Seasonal Variations in Total Catechins (TC) at Timbilil and Kipkebe

The total catechin content in the 20 clones presented in **Table 6**, varied significantly across the four environments. Considering the overal means, clone TRFK 303/ 1199 had significantly higher content of catechins at 19.37%, while clone TRFK 306 had the least catechins content at 10.02%. However, clone TRFK 12/19, TRFK 303/577, TRFK 12/12 and TRFK 11/4 had similar higher catechin content.

At the Timbilil site, clone TRFK 371/3 had the highest total catechins at 18.85%, while clone TRFK 306 had the least at 9.23% in wet season. In the dry season clone TRFK 6/10 had the highest catechins content at 18.05%, while clone TRFK 306 had the least at 10.63%. Although most of the clones showed significant decrease in total catechins during dry season, it is worth noting that clones TRFK 100/5, TRFK 6/8, and TRFK 12/19 showed a more decrease at the onset of water stress.

At the kipkebe site, clone TRFK 12/19 had the highest level of catechins at 20.27%, while clone TRFK 306 had the least at 10.27%, during wet season. During dry season, clone TRFK 303/1199 had the highest catechin content at 20.11%, while TRFK 306 had the least at 9.94%. The drastic decrease in total catechin content for clones TRFK 12/19, TRFK 11/4, TRFK 7/3 and TRFK12/12 during dry season is notable. Conversly, clone TRFK 430/90 showed a considerable increase in total catechins due to water stress. Thus, the clones were not affected in the same way by the changes in weather.

3.1.3. Seasonal Variations in Gallic Acid (GA) at Timbilil and Kipkebe

Garlic Acid content among the 20 clones varied signifycantly ($p < 0.05$) (**Table 7**). The mean of Garlic acid varied across the four environments. In regards to the overall means, clone TRFK 301/4, had the highest GA content at 1.08%, while clone TRFK 6/10 had the least at 0.40%. Interestingly, Clones BBK 35 and TRFK 301/6; clones TRFK 301/5 and TRFK 12/19; clones TRFK 430/90 and TRFK 12/56; and clones TRFK 303/577, TRFK 100/5 and TRFK 31/8 had equal GA content.

At the Timbilil site wet season, clone TRFK 301/4 the highest GA content while, clone TRFK 12/19, had the least in both seasons (**Table 6**). There was a decline in GA content in dry season for all the clones except for clones TRFK 12/19, TRFK 430/90, TRFK 31/11, TRFK 100/5 and TRFK 371/3, which had higher GA during dry season. In Kipkebe, clone TRFK 301/4 had the highest GA content, while clones TRFK 6/10 and TRFK 371/3 had the least in both seasons. There was significant decline in GA content as a result of water stress for all the clones except for clones TRFK 6/8, BBK35 and TRFK 31/11 which had significant increase. It is noteworthy that clones TRFK 430/90 and TRFK 371/3 were not affected by the changes in environment.

3.1.4. Seasonal Variations in Epigallocatechin (EGC) at Timbilil and Kipkebe

EGC content varied significantly with sites and seasons for the 20 clones presented in **Table 8**. Considering the overal mean clone TRFK 430/90 had the highest EGC content, while clone TRFK 306 had the least. Clones TRFK 6/8 and TRFK 301/5 had notably higher EGC content. At Timbilil site, wet season, clone TRFK 6/8 had the highest EGC content at 6.69%, while clone TRFK306 had the least at 1.94%. In the dry season, clone TRFK 430/90 had the highest EGC content at 6.36%,while clone TRFK 31/11 had the least at 2.63%. All the clones except for clones TRFK 301/6, TRFK 303/577, TRFK

12/19, TRFK 11/4, TRFK 6/8 and TRFK 371/3 had a increase in EGC content during dry season.

Table 5. Seasonal variations of total polyphenols (TP) at the Timbilil and Kipkebe sites.

Clone	Timbilil		Kipkebe		Overall means
	Wet season	Dry season	Wet season	Dry season	
TRFK 6/10	21.05	19.04	20.04	21.45	20.40
TRFK 301/6	21.48	18.32	19.81	20.94	20.14
TRFK 303/577	21.86	20.17	21.8	22.53	21.59
TRFK 12/19	22.21	22.14	22.99	25.09	23.11
TRFK 306	21.94	21.79	22.02	20.94	21.67
TRFK 301/5	21.39	19.83	20.18	21.34	20.68
TRFK 12/12	24.97	22.89	25.17	25.16	24.55
TRFK 11/4	20.74	20.92	22.99	23.03	21.92
TRFK 301/4	21.61	23.84	23.21	21.57	22.56
TRFK 430/90	22.85	20.69	23.26	23.43	22.56
TRFK 12/56	23.78	21.42	21.91	22.52	22.41
TRFK 7/3	22.76	19.26	22.42	23.65	22.02
TRFK 100/5	22.19	19.6	23.54	26.13	22.86
EPK C12	19.81	21.93	21.02	22.22	21.24
TRFK 303/1199	22.46	23.14	23.17	21.06	22.46
TRFK 6/8	25.09	19.09	24.8	23.91	23.22
BBK 35	25.52	22.04	22.91	24.77	23.81
TRFK 31/8	24.34	22.26	24.67	23.59	23.71
TRFK 31/11	25.3	22.08	21.82	26.89	24.02
TRFK 371/3	26.49	19.66	23.32	26.23	23.92
Overall mean	22.89	21.01	22.55	23.32	22.44
LSD (p = 0.05)	1.26	1.98	4.93	2.1	
C.V (%)	3.29	17.01	1.73	5.4	

Table 6. Seasonal variations in total catechins (TC) at Timbilil and Kipkebe sites.

Clone	Timbilil		Kipkebe		Overall means
	Wet season	Dry season	Wet season	Dry season	
TRFK 6/10	16.94	18.05	18.65	18.26	17.98
TRFK 301/6	16.98	15.93	16.53	17.42	16.71
TRFK 303/577	18.91	17.21	17.38	18.87	18.09
TRFK 12/19	18.57	15.74	20.27	18.02	18.15
TRFK 306	9.23	10.63	10.27	9.94	10.02
TRFK 301/5	16.93	16.91	19.75	18.1	17.92
TRFK 12/12	18.83	17.17	20.17	17.59	18.44
TRFK 11/4	19.1	17.49	19.88	17.79	18.56
TRFK 301/4	15.46	15.87	17.22	16.51	16.27
TRFK 430/90	16.83	17.67	17.3	19.38	17.80
TRFK 12/56	16.14	17.19	17.98	16.81	17.03
TRFK 7/3	16.81	17.86	17.18	15.77	16.91
TRFK 100/5	15.51	18.04	17.97	18.14	17.41
EPK C12	14.97	17.87	17.67	18.67	17.29
TRFK 303/1199	18.79	19.27	19.32	20.11	19.37
TRFK 6/8	18.46	15.9	15.75	16.41	16.63
BBK 35	17.62	16.7	18.37	18.99	17.92
TRFK 31/8	17.19	16.54	18.92	18.98	17.91
TRFK 31/11	14.41	16.24	18.99	19.77	17.35
TRFK 371/3	18.85	17.22	17.08	16.99	17.53
Overall mean	16.83	16.78	17.83	17.63	17.27
LSD ($p < 0.05$)	1.64	1.70	2.17	1.77	
C.V (%)	5.85	6.07	7.30	6.03	

At Kipkebe site, clone TRFK 301/5 had the highest EGC content at 6.76%, while clone TRFK 306 had the least at 1.26% during the wet season. In the dry season, clone TRFK 303/577 had the highest EGC content at 6.18% while clone TRFK 306 had the least at 2.59%. although most of the clones had an increase in EGC content in dry season, clone TRFK 306 and TRFK 430/90 had a drastic increase during dry season.

3.1.5. Seasonal Variations in Catechin (C) at Timbilil and Kipkebe

As shown in **Table 9**, catechin content varied significantly between sites and seasons except for Kipkebe wet season. Considering the means, clone TRFK 306 had the highest catechin content at 0.57%, while clone TRFK 31/11 had the least at 0.26%. Clones TRFK 6/10 and TRFK 301/6; clones TRFK 303/577 and TRFK 12/19; clones EPK C12 and TRFK 11/4; and clones TRFK 7/3, TRFK100/5 and TRFK 430/90 had equal catechin contents. At the Timbilil site, wet season, clone TRFK 12/12 had the highest catechin (+C) at 0.72%, while clone 31/11 had the least at 0.12%. Conversely during the dry season, clone TRFK 306 had the highest (+C) at 0.74% while clone TRFK 31/11 had the least at 0.17%. At the Kipkebe site, wet season, clone TRFK 303/577, had the highest catechin at 0.71%, while clone TRFK 301/5 had the least at 0.18%. In dry season, clone TRFK 306 had the highest catechin content, while clone TRFK 301/6 had the least at 0.22%. It is worth noting that most of the clones had lower catechins during dry season. However, clone TRFK 306 had considerable increase in catechin content during the dry season for both sites.

3.1.6. Seasonal Variations in Caffeine at Timbilil and Kipkebe

The level of caffeine varied considerably among the 20 clones presented in **Table 10**. Taking into account the overall means, clone TRFK 12/12 had the highest caffeine content at 4.11%, while clone TRFK 306 had the least at 1.94%. At the Timbilil site, clone TRFK 371/3 had the highest caffeine level at 3.93%, while clone TRFK 306 had the least at 1.76% during wet season. In the dry season, clone TRFK 6/10 had the highest caffeine content at 4.53%, while clone TRFK 306 had the least at 2.10%. At the Kipkebe site, wet season, clone TRFK 12/12 had the higest caffeine content at 3.25% while clone TRFK 306 and TRFK 303/577 had the least at 1.91%. Majority of the clones had a significant increase in caffeine level as a result of water stress in both sites as shown in **Table 9**. However, clone TRFK 303/577 and TRFK 12/12 had a drastic increase in caffeine level at the Kipkebe site. It is particularly worth noting that the caffeine content generally rose by 15.6% and 44.8% from wet to dry seasons for Timbilil and Kipkebe respectively.

Table 7. Seasonal variations in garlic acid (GA) at Timbilil and Kipkebe sites.

Clone	Timbilil		Kipkebe		Overall means
	Wet season	Dry season	Wet season	Dry season	
TRFK 6/10	0.49	0.36	0.39	0.36	0.40
TRFK 301/6	0.78	0.48	0.67	0.52	0.61
TRFK 303/577	0.62	0.44	0.59	0.44	0.52
TRFK 12/19	0.37	0.37	0.56	0.55	0.46
TRFK 306	0.82	0.73	1.03	0.79	0.84
TRFK 301/5	0.54	0.38	0.57	0.37	0.46
TRFK 12/12	0.62	0.58	0.73	0.69	0.66
TRFK 11/4	0.62	0.51	0.79	0.66	0.65
TRFK 301/4	1.08	1.01	1.11	1.1	1.08
TRFK 430/90	0.4	0.42	0.47	0.47	0.44
TRFK 12/56	0.37	0.41	0.53	0.47	0.44
TRFK 7/3	0.47	0.38	0.46	0.42	0.43
TRFK 100/5	0.46	0.52	0.55	0.54	0.52
EPK C12	0.54	0.48	0.67	0.59	0.57
TRFK 303/1199	0.59	0.43	0.68	0.51	0.55
TRFK 6/8	0.53	0.43	0.41	0.53	0.48
BBK 35	0.65	0.51	0.62	0.67	0.61
TRFK 31/8	0.52	0.39	0.61	0.56	0.52
TRFK 31/11	0.54	0.61	0.91	0.99	0.76
TRFK 371/3	0.4	0.64	0.38	0.38	0.45
Overall mean	0.57	0.5	0.64	0.58	0.57
C.V (%)	16.27	17.01	13.66	11.78	
LSD	0.15	0.14	0.14	0.11	

Table 8. Seasonal variations in epigallocatechin (EGC) at Timbilil and Kipkebe sites.

Clone	Timbilil		Kipkebe		Overall means
	Wet season	Dry season	Wet season	Dry season	
TRFK 6/10	5.48	6.04	6.11	6.13	5.94
TRFK 301/6	3.11	3.06	2.89	2.95	3.00
TRFK 303/577	6.10	6.08	5.12	6.18	5.87
TRFK 12/19	5.53	4.96	5.23	4.78	5.13
TRFK 306	1.94	2.69	1.26	2.59	2.12
TRFK 301/5	5.47	5.77	6.76	6.17	6.04
TRFK 12/12	5.19	4.73	5.46	4.29	4.92
TRFK 11/4	6.23	5.89	4.88	5.52	5.63
TRFK 301/4	2.74	2.98	3.69	3.24	3.16
TRFK 430/90	5.68	6.36	5.58	6.59	6.05
TRFK 12/56	4.06	5.59	4.73	4.79	4.79
TRFK 7/3	4.77	6.11	5.51	4.97	5.34
TRFK 100/5	4.70	5.84	5.67	5.93	5.53
EPK C12	4.14	5.52	4.62	5.30	4.90
TRFK 303/1199	5.08	5.56	4.66	5.30	5.15
TRFK 6/8	6.69	5.98	5.85	5.44	5.99
BBK 35	4.18	4.89	5.05	5.42	4.88
TRFK 31/8	5.27	5.77	5.87	5.79	5.67
TRFK 31/11	2.01	2.63	3.94	4.13	3.18
TRFK 371/3	5.42	4.49	5.40	5.37	5.17
Grand mean	4.69	5.05	4.91	5.04	4.92
C.V (%)	8.47	10.43	13.33	7.28	
LSD ($p < 0.05$)	0.66	0.88	1.09	0.61	

3.1.7. Seasonal Variations in Epicatechin (EC) at Timbilil and Kipkebe

There was a substantial variation ($p < 0.05$) in EC content among the 20 clones presented in **Table 11**. With reference to the overall means, clone TRFK 301/6 had the highest EC content at 4.15%, while clone TRFK 31/8 had the least at 1.14%. Clones TRFK 430/90 and TRFK 371/3 had equal EC content. Furthermore, the Cambod clones TRFK 301/5 and TRFK 301/4 had notably higher EC than the other clones. At the Timbilil site, during the wet season, clone TRFK 301/6 had the highest EC content at 4.65%, while clone TRFK 306 had the least at 1.00%. In the dry season clone TRFK 301/6 still had the highest EC content at 4.56%, while clone TRFK 11/4 had the least at 1.10%. Though most of the clones had lower EC content in dry season, clones TRFK 306, TRFK 301/4, TFK 12/56, TRFK 7/3 and EPKC12 had higher EC in the dry season. At kipkebe site, during wet season, clone TRFK 301/5 had the highest level of EC at 4.86%, while clone TRFK 31/11 had the least EC content at 0.89%. On the other hand, during the dry season, clone TRFK 301/6 had the highest EC content at 4.91%, while clone TRFK31/8 had the least at 1.02%. At this site, most of the clones had an increase in EC content. It is worth noting that at the two sites, Cambod clones TRFK301/5, TRFK 301/4 and TRFK 301/6 generally had high EC content than the Assam clones.

3.1.8. Seasonal Variations in Epigallocatechin Gallate (EGCG) at Timbilil and Kipkebe

EGCG content in the 20 clones presented in **Table 12** varied considerably across four environments. Taking into account the overall means, clones TRFK 303/1199 had the highest EGCG content at 9.01%, while clone TRFK 306 had the least at 3.74%. Clones TRFK 12/12 and clone TRFK 31/11 had equal mean EGCG contents. At the Timbilil site, wet season, clone TRFK 11/4 had the highest EGCG content at 8.86%, while clone TRFK 306 had the least at 3.34%. In the dry season, clone TRFK 371/3 had the highest EGCG content at 8.78%, while clone TRFK 301/6 had the least at 3.25%. Most of the clones had a decline in their EGCG content during the dry season, although clones TRFK 6/8 and TRFK 31/11 showed more decrease. Nevertheless, at the Kipkebe site, clone TRFK 31/11 had the highest EGCG content at 10.38%, while clone TRFK 6/8 had the least at 5.93% during wet season. During the dry season, clone 31/11 had the highest EGCG content at 10.68%, while clone TRFK 301/6 had the least at 3.33%. Majority of the clones had EGCG content decline during the dry season.

Table 9. Seasonal variations in catechin (C) at Timbilil and Kipkebe sites.

Clone	Timbilil		Kipkebe		Overall means
	Wet season	Dry season	Wet season	Dry season	
TRFK 6/10	0.21	0.30	0.36	0.38	0.31
TRFK 301/6	0.54	0.18	0.31	0.22	0.31
TRFK 303/577	0.30	0.35	0.71	0.34	0.43
TRFK 12/19	0.56	0.31	0.52	0.31	0.43
TRFK 306	0.40	0.74	0.44	0.69	0.57
TRFK 301/5	0.39	0.26	0.18	0.27	0.28
TRFK 12/12	0.72	0.36	0.54	0.38	0.5
TRFK 11/4	0.35	0.28	0.43	0.27	0.34
TRFK 301/4	0.46	0.27	0.34	0.25	0.33
TRFK 430/90	0.50	0.28	0.32	0.33	0.36
TRFK 12/56	0.29	0.31	0.39	0.3	0.32
TRFK 7/3	0.34	0.34	0.48	0.29	0.36
TRFK 100/5	0.34	0.34	0.42	0.35	0.36
EPK C12	0.25	0.34	0.41	0.35	0.34
TRFK 303/1199	0.14	0.29	0.38	0.36	0.29
TRFK 6/8	0.26	0.41	0.51	0.46	0.41
BBK35	0.33	0.32	0.46	0.36	0.37
TRFK 31/8	0.16	0.33	0.33	0.34	0.29
TRFK 31/11	0.12	0.17	0.4	0.35	0.26
TRFK 371/3	0.49	0.30	0.56	0.56	0.47
Grand mean	0.36	0.32	0.42	0.36	0.37
C.V (%)	19.44	10.93	41.61	9.47	
LSD	0.12	0.06	NS	0.06	

NS: denotes non-significant.

Table 10. Seasonal variations in caffeine (Caff) at Timbilil and Kipkebe sites.

Clone	Timbilil		Kipkebe		Overall means
	Wet season	Dry season	Wet season	Dry season	
TRFK 6/10	3.86	4.53	2.43	4.3	3.78
TRFK 301/6	2.61	3.11	1.94	2.63	2.57
TRFK 303/577	2.79	3.18	1.91	4.11	3.00
TRFK 12/19	3.32	3.50	2.96	3.76	3.38
TRFK 306	1.76	2.10	1.91	1.97	1.94
TRFK 301/5	3.15	4.46	2.78	3.95	3.59
TRFK 12/12	3.87	4.73	3.25	4.6	4.11
TRFK 11/4	3.53	4.04	3.02	3.43	3.50
TRFK 301/4	2.58	3.4	2.07	2.86	2.72
TRFK 430/90	2.83	3.23	2.00	3.34	2.85
TRFK 12/56	3.17	3.72	2.5	3.64	3.26
TRFK 7/3	3.17	3.21	2.03	3.22	2.91
TRFK 100/5	3.06	4.04	2.44	3.24	3.2
EPK C12	2.95	3.92	2.38	3.52	3.19
TRFK 303/1199	3.35	3.39	2.7	3.6	3.26
TRFK 6/8	2.63	2.91	1.94	3.11	2.65
BBK 35	3.34	3.9	2.37	3.65	3.32
TRFK 31/8	3.45	3.68	2.55	3.73	3.35
TRFK 31/11	2.34	2.7	2.21	3.35	2.65
TRFK 371/3	3.93	3.48	2.78	3.69	3.47
Overall mean	3.08	3.56	2.41	3.49	3.13
C.V (%)	8.14	10.14	6.14	8.59	
LSD ($p < 0.05$)	0.42	0.6	0.35	0.5	

Table 11. Seasonal variations in total Epicatechins (EC) at Timbilil and Kipkebe sites.

Clone	Timbilil		Kipkebe		Overall means
	Wet season	Dry season	Wet season	Dry season	
TRFK 6/10	1.25	1.13	2.03	1.85	1.56
TRFK 301/6	4.65	4.56	2.5	4.91	4.15
TRFK 303/577	1.87	1.82	1.55	1.21	1.61
TRFK 12/19	1.55	1.30	2.02	1.29	1.54
TRFK 306	1.00	1.37	1.30	1.25	1.23
TRFK 301/5	3.28	2.96	4.86	3.55	3.66
TRFK 12/12	1.94	1.71	2.78	1.73	2.04
TRFK 11/4	1.33	1.1	1.88	1.11	1.35
TRFK 301/4	2.94	3.29	2.11	2.92	2.82
TRFK 430/90	1.62	1.48	1.33	1.67	1.52
TRFK 12/56	1.36	1.43	1.06	1.28	1.28
TRFK 7/3	1.78	1.97	1.46	1.45	1.66
TRFK 100/5	1.38	1.33	1.22	1.44	1.34
EPKC 12	1.72	1.87	1.44	1.54	1.64
TRFK 303/1199	1.73	1.67	1.36	1.67	1.61
TRFK 6/8	2.05	1.69	1.52	1.54	1.7
BBK 35	1.89	1.47	1.41	1.60	1.59
TRFK 31/8	1.34	1.20	1.02	1.02	1.14
TRFK 31/11	3.74	2.97	0.89	1.10	2.18
TRFK 371/3	2.07	1.11	1.63	1.62	1.61
Overall mean	2.02	1.87	1.77	1.79	1.86
C.V (%)	10.4	21.82	24.5	14.98	
LSD ($p < 0.05$)	0.35	0.68	0.72	0.45	

3.1.9. Seasonal Variations in Epicatechin Gallate (ECG) at Timbilil and Kipkebe

There was significant variation in ECG content in the 20 clones presented in **Table 13**. In view of the overall means, clone TRFK 301/6 had significantly the highest ECG content at 5.03%, while clone TRFK 100/5 had the least at 1.96%. Clones TRFK 303/577 and TRFK 306 had equal ECG content. At the Timbilil site clone TRFK 301/6 had the highest ECG content at 5.14%, while clones TRFK 430/90 had the least at 1.87% during the wet season. In dry season, clone TRFK 301/6 had the highest ECG at 4.88%, while clone TRFK 31/8 had the least at 1.82%. Most of the clones in this site had lower ECG content in the dry season. However, at the Kipkebe site wet season, clone TRFK 301/4 had the highest ECG content at 4.71%, while clone TRFK 6/8 had the least at 1.93%. In the dry season, clone TRFK 301/6 had the highest ECG content at 6.02% while clone TRFK 100/5 had the least at 1.66%. Cambod varieties had higher level of digallated catechins such as EC and ECG.

3.2. Principle Component Analysis for the 20 Clones Using Mean Catechin Data across Two Sites and Seasons

The principle component analysis grouped the clones into three clusters (**Figure 2**). The first principle component accounted for 58.72% of the total variation, while the second principle component accounted for 21% variation, with an accumulation of 79.72% variation.

3.3. PCA for the Total Polyphenols for 20 Clones Using Mean Polyphenol Contents across Four Environments

The principle component analysis on the total polyphenols presented in **Figure 3** similarly, revealed three distinct clusters. The first principle component accounted for 79.85% of the total variation, while the second principle component accounted for 11.03% variation. The cumulative variation was 90.88%.

Table 12. Seasonal variations in Epigallocatechin gallate (EGCG) at Timbilil and Kipkebe sites.

Clone	Timbilil		Kipkebe		Overall means
	Wet season	Dry season	Wet season	Dry season	
TRFK 6/10	8.04	8.64	7.79	7.25	7.93
TRFK 301/6	3.54	3.25	6.79	3.33	4.23
TRFK 303/577	8.07	6.92	7.41	8.89	7.82
TRFK 12/19	7.85	7.06	9.51	8.89	8.33
TRFK 306	3.34	3.60	4.35	3.67	3.74
TRFK 301/5	4.57	4.99	4.78	4.79	4.79
TRFK 12/12	7.89	7.44	8.36	7.84	7.88
TRFK 11/4	8.86	8.25	10.08	8.66	8.96
TRFK 301/4	4.27	4.51	6.38	4.98	5.04
TRFK 430/90	7.16	7.46	7.63	8.31	7.64
TRFK 12/56	8.05	7.65	9.25	7.99	8.24
TRFK 7/3	7.52	6.94	7.34	6.82	7.16
TRFK 100/5	7.22	8.38	8.49	8.77	8.21
EPK C12	6.35	7.45	8.16	8.64	7.65
TRFK 303/1199	8.79	8.71	9.22	9.33	9.01
TRFK 6/8	7.55	5.91	5.93	6.78	6.54
BBK 35	8.18	7.22	8.29	8.27	7.99
TRFK 31/8	8.48	7.42	9.45	9.30	8.66
TRFK 31/11	4.3	6.15	10.38	10.68	7.88
TRFK 371/3	8.73	8.78	6.86	6.82	7.8
Overall mean	6.94	6.84	7.82	7.5	7.27
C.V (%)	9.28	11.28	13.08	7.79	
LSD ($p < 0.05$)	1.07	1.29	1.70	0.97	

Table 13. Seasonal variations in Epicatechin gallate (ECG) at Timbilil and Kipkebe sites.

Clone	Timbilil		Kipkebe		Overall means
	Wet season	Dry season	Wet season	Dry season	
TRFK 6/10	1.97	1.95	2.37	2.65	2.23
TRFK 301/6	5.14	4.88	4.04	6.02	5.02
TRFK 303/577	2.56	2.03	2.6	2.25	2.36
TRFK 12/19	3.09	2.12	2.98	2.73	2.73
TRFK 306	2.55	2.23	2.92	1.73	2.36
TRFK 301/5	3.21	2.92	3.16	3.32	3.15
TRFK 12/12	3.09	2.93	3.04	3.35	3.1
TRFK 11/4	2.33	1.97	2.61	2.23	2.28
TRFK 301/4	5.04	4.83	4.71	5.11	4.92
TRFK 430/90	1.87	2.1	2.45	2.47	2.22
TRFK 12/56	2.39	2.22	2.55	2.44	2.4
TRFK 7/3	2.4	2.5	2.39	2.23	2.38
TRFK 100/5	1.87	2.15	2.17	1.66	1.96
EPK C12	2.51	2.68	3.02	2.84	2.76
TRFK303/1199	3.04	3.05	3.69	3.46	3.31
TRFK 6/8	1.92	1.91	1.93	2.19	1.99
BBK 35	3.04	2.8	3.16	3.33	3.08
TRFK 31/8	1.94	1.82	2.25	2.53	2.14
TRFK 31/11	4.23	4.31	3.37	3.51	3.86
TRFK 371/3	2.15	2.54	2.63	2.62	2.48
Overall mean	2.82	2.7	2.9	2.93	2.84
C.V (%)	9.31	8.96	17.12	14.76	
LSD ($p < 0.05$)	0.44	0.4	0.83	0.72	

4. DISCUSSION

The results from catechin analysis indicate that the level of biochemical components in tea is varietal dependent. This is an indication that each individual clone is unique in the level of catechins. Most of the clones used showed some level of association between the total polyphenols and soil water content. There was lower polyphenols content in dry season compared to the wet season in most of the clones. This was expected because water is one of the raw materials for photosynthesis and it has direct impacts on organic synthesis of plants of both the primary and the secondary metabolites. The leaf is the major source of photo assimilates, which yields the major precursors of secondary metabolism such as malonylCoA and coumaroyl-CoA which are dependent on the light and the soil water content [10]. Previous results showed that phenolics in green tea shoots varied among clones [17], while other results revealed correlation between water content and biochemical parameters [13].

The clones had high level of caffeine during the dry season. This is because accumulation of secondary metabolites such as caffeine by plants is known to be a form of defence mechanism adopted by plants. This mechanism helps plants respond to and adapt to environmental stresses [18]. Water stress is one of the most important environmental factors that can regulate plant growth and development, limit plant production, and alter the physiological and biochemical properties of plants. Indeed, water stress is known to increase the amount of secondary metabolites in plants [19]. One of the responses to abiotic stress is the accumulation of non-toxic, low molecular weight compounds, called compatible solutes or osmolytes, such as caffeine. In response to abioitc stress there is increase in the level of caffeine during dry season. From the this study, it was observed that all the clones except for clone TRFK 306 increased significantly in their level of caffeine as a result of the dry season. The accumulation of caffeine could be due to the hydroxylation of p-ce toumaric acid to caffeic acid in the phenylpropanoid pathway through the influence of polyphenol oxidase (PPO's) [20]. This could be the reason for the increase in the level of caffeine in the tea clones as a result of the dry season. Expression analysis done showed that polyphenol oxidase (PPO) was induced in response to water stress with the highest magnitude of induction in older leaves and corresponding abscission zones, which might facilitate cell death preferentially in these tissues [21].

The new improved clones had high levels of catechins like the other commercialized clones used in this study except for clone TRFK 306. The low levels of catechins in fresh leaves of clone TRFK 306 could be attributed to the up regulation of the anthocyanin synthase in biosynthetic pathway for anthocyanins rather than the leucoanthocyanin reductase in catechin biosynthetic pathway probably due to environmental stimuli [22]. This could imply that this variety clones could be commercialized for the production of purple tea which has been considered a medicine and a healthful beverage [23]. Furthermore, some clones had high levels of catechins, especially EGCG, which from the preceding studies done indicated that such high levels of EGCG are important health benefits [24].

The catechin content in some of the clones was not significantly affected by seasonal variation. This could be attributed to the slight changes in the catechin levels in individual clones, resulting from seasonal variation due to the principal amounts of the specific catechins in the leaf. However, organic compounds that are synthesized in small quantities, such as C, ECG, EGC and EC, could be expected to exhibit significant changes when the plant is under stress, but this was not the case for the results of some of the clones from this study. This observation rules out the influence of principal amount of catechins in plants as the basis of the differential water stress response of the plants. The results obtained from this study are corroborated by earlier result which indicated that the changes in soil water content did not have effects on the level of catechins [13].

On the other hand, the differences in the levels of catechins in the different clones could be attributed to the up regulation or down regulation of the enzyme flavanone 3-hydroxylase. Flavanone 3-hydroxylase (F3H) is an abundant enzyme in tea leaves that catalyses the stereo specific hydroxylation of (2S)-naringenin to form (2R, 3R)-dihydrokaempferol. In a study carried out to determine the concentration of catechins, and CsF3H gene expression found a positive correlation between the concentration of catechins, and CsF3H gene expression in leaves at different developmental stage. CsF3H expression was down-regulated in response to drought, abscisic acid and gibberellic acid treatment, but up-regulated in response to wounding. The concentration of catechins paralleled the expression data. In addition, they also reported that drought decreased the level of catechins in tea [13]. The decrease in catechins can be exemplified by percentage total catechins in clones TRFK 12/19, TRFK 12/12, and TRFK 11/4 during the season. In a recent study high flavanone-3-hydroxylase (F3H) activity in tea leaves was reported suggesting it as an important enzyme in the biosynthesis of catechins in tea [22]. These data indicate that catechin may serve as a feedback inhibitor of CsF3H expression [25]. Negative feedback regulation of CsF3H expression by catechins could explain the storage of catechins in the vacuole which would allow for their continuous synthesis [26]. This could explain the lack of change in the level of total catechins in some of the clones such as TRFK 371/3, TRFK 301/4, and TRFK 6/10 in the four environments.

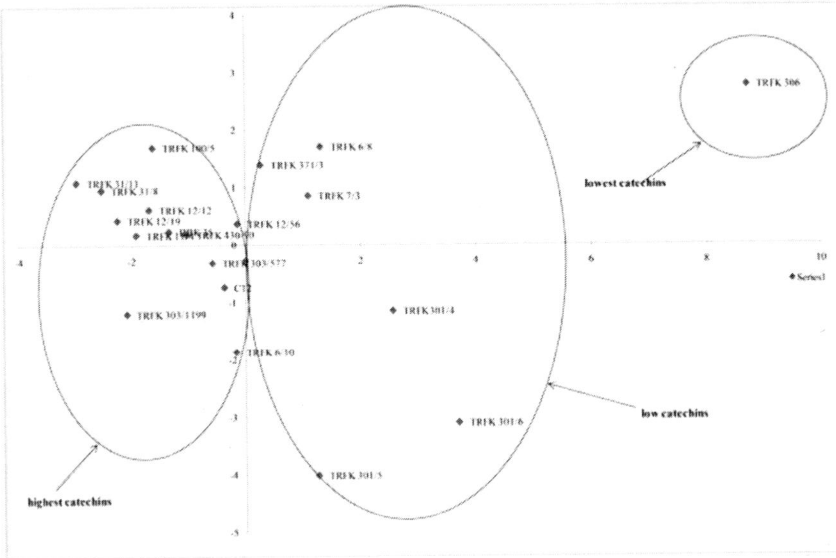

Figure 2. Principle component analysis for the 20 clones using mean catechin data across sites and seasons.

From principle component analysis, a clear structure was observed within the different clones of tea with the identification of different clusters based on the catechins and polyphenols shown in Figures 2 and 3. The Assam teas have higher catechins content compared to the Cambod teas, which have moderate catechins. This could be attributed to the unique catechin profile of Cambod varieties. The Cambod clones were low in EGCG, medium low in EGC and high in EC and ECG content. These results are similar to the results obtained in a preceding study [6]. Catechin rich clones could be distinguished from the low catechins clones within the PCA plot. Thus individual catechins can be used in the identification and selection of high quality clones.

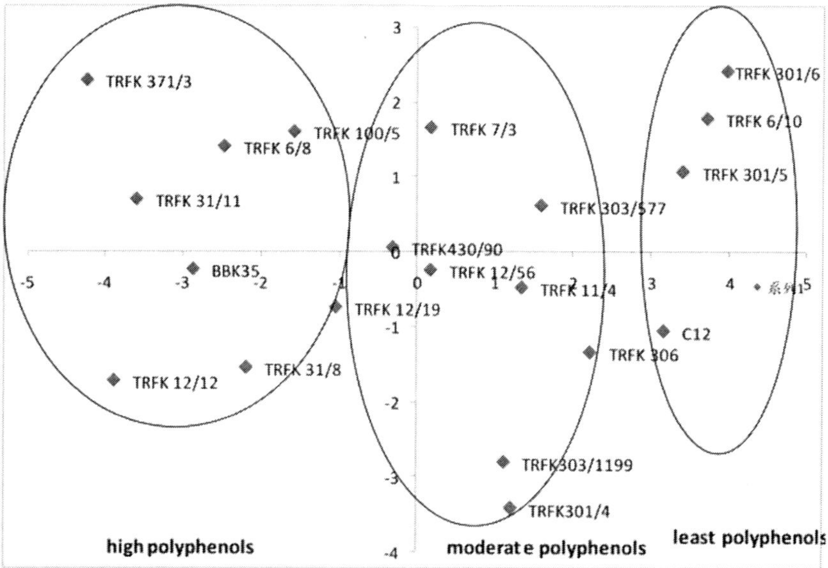

Figure 3. Scatter graph PCA for total polyphenols based on the mean polyphenol content across seasons and sites.

5. CONCLUSION

From this study it can be concluded that the level of catechins in tea plants are varietal dependent. Environmental stress has effects on the synthesis of Flavanoids as well as the levels of caffeine in plants. The results presented indicated that the differences in the accumulation of catechins among the tea clones, and by extension important for the discrimination of clonal diversity. The determination of the influence of wider climatic conditions on the biosynthesis and the content of Flavanoids could be explored as the subjects for further research.

REFERENCES

1. M. Obanda, P. O. Owuor and C. K. Njuguna, "The Impact of Clonal Variation of Total Polyphenols Content and Polyphenol Oxidase Activity of Fresh Tea Shoots on Plain Black Tea Quality Parameters," Tea, Vol. 13, No. 2, 1992, pp. 129-133.

2. H. Wang and K. Helliwell, "Determination of Flavanols in Green Tea and Black Tea Leaves and Green Tea Infusions by High Perfromance Liquid Chromatography," Food Reserves International, Vol. 34, No. 2-3, 2001, pp. 223-227.

3. Y. Hara, "Prophylactic Functions of Tea Catechins," In: W. K. Jain, M. A. Siddiqi and J. H. Weisburger, Eds., Protective Effects of Tea on Human Health, CABI Publishing, Wallingford, 2006, pp. 16-24.

4. M. Boschmann and F. Thielecke, The Effects of Epigallocatechin-3-Gallate on Thermogenesis and Fat Oxidation in Obese Men: A Pilot Study," Journal of the American College of Nutrition, Vol. 26, No. 4, 2007, pp. 389-395.

5. A. Persson, M. Josefsson, K. Persson and R. G. Andersson, "Tea Flavonols Inhibit Angiotensin-Converting Enzyme Activity and Increase Nitric Oxide Production in Human Endothelial Cells," Journal of pharmacy and Pharmacology, Vol. 58, No. 8, 2007, pp. 1139-1144.

6. G. Goodarznia and A. Abdollahi, "Superheated Water Extraction of Catechins from Green Tea Leaves: Modelling and Simulation," Chemistry and Chemical Engineering, Vol. 16, No. 2, 2009, pp. 99-107.

7. Tea Research Foundation of Kenya, "Tea Cultivation Manual for Good Agricultural Practices," Inkit, Nairobi, 2012, p. 70.

8. S. A. Laurie, V. A. Miller, S. C. Grant and M. G. Kris, "Phase I Study of Green Tea Extract in Patients with Advanced Lung Cancer," Cancer and Chemotherapy Pharmacology, Vol. 55, No. 1, 2005, pp. 33-38.

9. F. N. Wachira, J. Tanaka and Y. Takeda, "Genetic Variation and Differentiation in Tea (Camellia Sinensis) Germplasm Revealed by RAPD and AFLP Variation," Journal of Horticultural Sciences and Biotechnology, Vol. 76, No. 5, 2001, pp. 557-563.

10. G. N. Magoma, et al., "The Use of Catechins as Biochemical Markers in Diversity Studies of Tea (Camellia Sinensis)," Genetic Resources and Crop Evolution, Vol. 47, No. 2, 2000, pp. 107-114.

11. D. A. Balentine, S. A. Wiseman and L. C. Bouwens, "The Chemistry of Tea Flavonoids," Food Science and Nutrition, Vol. 37, No. 8, 1997, pp. 693-704.

12. K. Singh, et al., "An Early Gene of the Flavonoid Pathway, Flavonone 3-Hydroxylase, Exhibits a Positive Relationship with the Concentration of Catechins in Tea (Camellia Sinensis)," Tree Physiology, Vol. 28, No. 9, 2008, pp. 1349-1356.

13. E. K. Cheruiyot, et al., "Polyphenols as Potential Indicators for Tolerance in Tea (Camellia Sinensis)," Bioscience, Biotechnology, and Biochemistry, Vol. 71, No. 9, 2007, pp. 2190-2197.

14. H. P. Singh, S. D. Ravindranath and C. Singh, "Analysis of Tea Shoot Catechins: Spectrophotometric Quantification and Selective Visualization on Two Dimensional Paper Chromatograms Using Diazotized Sulphanilamide," Journal of Agricultural and Food Chemistry, Vol. 47, No. 3, 1999, pp. 1041-1045.

15. B. O. Kwach, et al., "Effects of Location on Production Fertilizer Rates and Plucking Intervals on Mature Leaf Nutrients of Clone TRFK 6/8 in East Africa," Tea, Vol. 32, No. 2, 2011, pp. 56-68.

16. Genstat 5 Committee, "Principle Component Analysis and Hierachial Ascendance Classification. In Genstat Reference Manual," Clarendon Press, Oxford, 1989, pp. 471-508.

17. M. Obanda and P. Owuor, "Flavonols Composition and Caffeine Content of Green Leaf as Quality Potential Indicators of Kenyan Black Teas," Journal of the Science of Food and Agriculture, Vol. 74, No. 2, 1997, pp. 209-215.

18. Kirakosyan, et al., "Applied Environmental Stresses to Enhance the Levels of Polyphenolics in Leaves of Hawthorn Plants," Physiologia Plantarum, Vol. 121, No. 2, 2004, pp. 182-186.

19. S. M. A. Zobayed, F. Afreen and T. Kozai, "Phytochemical and Physiological Changes in the Leaves of St. John's Wort Plants under a Water Stress Condition," Environmental and Experimental Botany, Vol. 59, No. 2, 2007 pp. 109-116.

20. G. Spagna, et al., "Characterization of a Tomato Polyphenol Oxidase and Its Role in Browning and Lycopene Content," Journal of Agricultural and Food Chemistry, Vol. 53, No. 6, 2005, pp. 2032-2038.

21. M. R. Badger, et al., "Electron Flow to Oxygen in Higher Plants and Algae: Rates and Control of Direct Photoreduction (Mehler Reaction)

and Rubisco Oxygenase," Philosophical Transactions of the Royal Society B, Vol. 355, No. 1402, 2000, pp. 1433-1446.

22. P. A. N. Punyasiri, et al., "Flavonoid Biosynthesis in the Tea Plant Camellia Sinensis: Properties of Enzymes of the Prominent Epicatechin and Catechin Pathways," Archives of Biochemistry and Biophysics, Vol. 431, No. 1, 2004, pp. 22-30.

23. C. Carmen, R. Artacho and R. Giménez, "Beneficial Effects of Green Tea-a Review," Journal of the American College of Nutrition, Vol. 25, No. 2, 2006, pp. 79-99.

24. G. Dale, et al., "Epigallocatechin-3-Gallate (EGCG): Chemical and Biomedical Perspectives," Phytochemistry, Vol. 67, No. 17, 2006, pp. 1849-1855.

25. L. Britsch and H. Grisebach, "Purification and Characterization of (2s)-Flavanone 3-Hydroxylase from Petunia Hybrida," European Journal of Biochemistry, Vol. 156, No. 3, 1986, pp. 569-577.

26. T. C. Fischer, H. Halbwirth and S. Roemmelt, "Nduction of Polyphenol Gene Expression in Apple (Malus × Domestica) after the Application of a Dioxygenase Inhibitor," Journal of Plant Physiology, Vol. 128, No. 4, 2006, pp. 604-617.

Chapter 11

Water Use and Drought Response in Cultivated and Wild Apples

Carole L. Bassett[1]

[1] Research Molecular Biologist, USDA, Agricultural Research Service, Appalachian Fruit Research Station, Kearneysville, WV, USA

1. INTRODUCTION

Water availability is the single most important factor determining plant survival. Many environments experience water-limited periods of various degrees and duration. The issue of water availability for agricultural crops where maintenance of high yields over variable growing seasons is desirable is particular critical. For plants, the strategy choices for survival can be summarized as dehydration avoidance (e.g., deep rooting), dehydration tolerance (e.g., accumulation of osmoprotectants) and drought escape (e.g., reproductive completion before the dry season). Drought adaptation in most plants is controlled by complicated interactions between anatomy, physiology and biochemistry, all of which are directly or indirectly under genetic control [1-3].

2. WATER USE EFFICIENCY AND DROUGHT RESISTANCE

All living organisms have evolved mechanisms for adapting to changes in their environment, whether biotic such as pest related, or abiotic such as physically effected. For plants this is especially challenging, since they are unable to relocate to avoid adverse conditions. As a result, numerous strategies employed by plants leading to successful acclimation have been identified. Broadly speaking, these adaptive mechanisms can be divided into two major categories, namely morphological modifications and physiological adjustments. Through combinations of these basic strategies, plants can respond quickly to environmental cues, often maintaining the response for relatively long periods [4,5].

The term drought resistance is sometimes confused with water use efficiency (WUE). Drought resistance is determined primarily by 'drought avoidance' (high plant water status maintained under water deficit) and/or 'drought tolerance' (capacity to sustain plant function in a dehydrated state) [6,7]. Drought resistance in a genetic/physiological context refers to the ability of one genotype to yield 'better' than another during severe drought stress. On the other hand WUE is defined as the ratio between diffusion of CO_2 into the leaf (photosynthesis) and loss of H_2O through transpiration, indicated as WUE = A/E, where A is carbon assimilation and E is transpiration. It is positively correlated with carbon isotope discrimination ($\Delta^{13}C$) based on the stable carbon isotope ratio, $\delta^{13}C$ ($^{12}C/^{13}C$ relative to a standard, i.e. PeeDee Belemnite), in the plant tissue relative to the atmospheric ratio and is calculated as: $\Delta^{13}C = \delta^{13}C$ in air - $\delta^{13}C$ of the plant/1- $\delta^{13}C$ of the plant. Since most gas exchange occurs via the stomata, it is expected that guard cell function would be closely associated with WUE. Indeed, the size and density of stomates correlates well with water use efficiency [8-10]. For drought resistance, yield is not necessarily adversely affected by resistance, whereas for WUE reduced transpiration through stomatal closure is often accompanied by reduced yield potential through reduced carbon assimilation, particularly in herbaceous C_3 plants (however, see below). Other parameters, such as root depth, leaf size, and trichome size and density have also been linked to water use efficiency [2], but they have also been linked to drought resistance as well [6].

Different methods have been used to measure drought resistance and WUE [11]. These methods measure the energy status of water in plant tissues and the trans-port processes into and out of the soil-plant-atmosphere continuum. In general, these methods isolate specific plant tissues at instantaneous moments in time, whereas $\Delta^{13}C$ represents a time-integrated value of the ratio of C_i (intercellular CO_2concentration) to ambient CO_2 which, as previously indicated, reflects the plant's capacity for gas exchange via stomata [12] and discrimination of rubisco against ^{13}C. The use of carbon isotope discrimination to select in-dividuals with higher WUE has been applied successfully to cereal breeding programs (13). The extent of $\delta\,^{13}C$ varies substantially among wheat genotypes, and heritability is high because genotype X environment interactions are relatively low [14, 15]. Rebetzke et al. [16] reported on the selection of plants with greater biomass, harvest index and kernel weight using results from contrasting high and low $\Delta^{13}C$ groups in combination with a backcrossing program. There were significant correlations between $\Delta^{13}C$ and yield and between $\Delta^{13}C$ and biomass. The resulting high yielding strain, 'Drysdale', produces around 10% more grain under drought conditions than other dry-area wheat varieties.

3. ADAPTATION AND THE RELATIONSHIP OF $\Delta^{13}C$ TO YIELD

Adaptive changes in populations growing in different environments have been amply demonstrated in a variety of plants [17]. Divergence among populations associated with different environments provides the raw material for speciation and differentiation among closely related species. Higher fitness of genotypes in their native environment compared to genotypes transplanted from contrasting environments provides evidence of local adaptation [9]. For example, when two populations of*Boechera holboellii* growing in xeric and wet environments were grown in reciprocal transplant experiments, significantly higher survival was observed with plants growing in their native habitat [18]. Furthermore, genes identified by cDNA-AFLP analysis showed genotype-specific expression patterns related to the indigenous environment (population). In another study populations of *Encelia farinosa*growing over a broad rainfall gradient in the deserts of southwestern North America, were evaluated for leaf characteristics. Leaf pubescence declined as mean water availability increased [19]. Variations among populations for both pubescence and carbon isotope discrimination persisted when the plants were grown in common environments differing in water availability, indicating a genetic basis for variation in these traits [20].

WUE in trees adapted to different environments has been documented in several forest species [21-24]. For example, a study of four birch (*Betula pendula* Roth) clones from environments with different rainfall amounts indicated strong clonal differences in a number of water use and photosynthetic traits, including leaf $\delta^{13}C$ values [21]. A follow-up study correlated differences in leaf and root morphological parameters and carbon partitioning with clones from the drier environments [25]. Although individual leaf areas were smaller in drought-treated clones, regardless of their region of origin, total leaf and specific leaf areas (leaf area/leaf weight) were actually higher for the drought-treated clones from the drier environment. This is in contrast to the often observed reduction in leaf surface area seen in plants exposed to water deficit (WD) [26].

Poplar species are differentially adapted to a variety of environments, and because poplar is a rapidly growing tree with heavy water use, there is growing interest in developing lines that are drought resistant. Links between productivity and $\Delta^{13}C$ varied in a study comparing different poplar genotypes, suggesting that genotypes displaying simultaneous high productivity and improved water use efficiency could be selected [27]. To obtain more practical information regarding productivity and WUE, a field study was conducted on the same genotypes analyzed in the previous study. Significant clonal diversity was observed for several traits related to productivity and for Δ which showed high heritability (H2 = 0.71) [28]. A lack of correlation between above ground biomass and Δ was reflected in several clones where high productivity was combined with improved WUE. This observation supports previous studies with cereals indicating that WUE and yield can be inherited as separate traits.

Yield of deciduous tree fruit crops is not measured as total biomass yield in commercial production as are agronomic crops such as corn, wheat and rice or forest trees. In commercial orchards it is common practice to reduce yield potential of fruit trees by as much as 50% to insure large fruit size and high fruit quality [29]. Consequently, the paradigm that increased WUE is tied to reduced yield potential is not necessarily valid for tree fruit production. For example, Glenn et al. [30] have demonstrated the practicality of identifying peach cultivars with high WUE without compromising productivity. This study, taken together with those cited previously, demonstrates the feasibility of selecting for improved WUE without loss in productivity.

4. ADAPTATION, WUE AND Δ ^{18}O

Transpiration rate (E) affects water loss to the atmosphere and is negatively correlated with WUE. Despite the fact that atmospheric ^{18}O is low (ca. 0.2% of total oxygen), plants tend to accumulate ^{18}O and ^{2}H in leaf water due to the difference in vapor pressure between heavy water and 'normal' water and to differences in diffusivity with air. However, the relationship between E and isotopic enrichment is complex. For example, variation in E can be caused by changes in evaporative demand and/or changes in stomatal conductance, gs [31]. If the source of variation is evaporative demand, then as E increases, ^{18}O enrichment increases. On the other hand, if stomates are the source of variation, then as E decreases (stomates close), leaf water enrichment increases. How does this relate to ^{18}O enrichment of plant organic matter? Plants accumulate ^{18}O in their tissues as a result of the exchange of oxygen between water and carbonyl oxygens in triose phosphates. An enrichment of about 27 parts per thousand (ppt) has been observed in the organic material of several different plants relative to leaf water [32]. This suggests that differences in ^{18}O enrichment can be used to distinguish genotypes with favorable yields and stomatal function. In fact Barbour et al. [33] recently demonstrated a reliably negative correlation between yield and $\delta^{18}O$ in wheat which was used to identify water use efficient varieties for breeding.

5. SPECIFIC GENES ASSOCIATED WITH WUE AND/OR DROUGHT RESPONSES IN APPLE AND OTHER PLANTS

The recent advent of global gene expression methodology has spawned a number of studies of abiotic stress responses, including drought, in several plant species [34-38]. In Arabidopsis, a compilation study of microarray analyses on plants subjected to a variety of stress treatments highlighted overlap among genes up-regulated in the early stages of all the stress responses [39]. Studies on WD stress in cereals and dicots have cataloged a large number of genes up-regulated during treatment [35-37, 40, 43]. Comparisons among these studies reveal that a number of genes are reproducibly up-regulated in response to WD regardless of how the stress was imposed or what plant system was involved, including apple (Table 1) [44].

Table 1. Genes up-regulated in water deficit-treated apple roots vs other plants responding to water deficit[1]

Genes Up-regulated in 'Royal Gala' Roots	Other Plants	Tissue	Citation
HMW HSP	*Arabidopsis*	various	[40][2]
	Poplar proteome	white roots	[41]
aquaporin	*Arabidopsis*	various	[40]
	barley	leaves and roots	[35]
	maize	leaves and roots	[43]
	Poplar proteome	white roots	[41]
protease inhibitor	*Arabidopsis*	various	[40,42]
	chickpea	whole seedlings	
Histone H2	*Arabidopsis*	various	[40]
	chickpea	whole seedlings	[42]
	maize	leaves and roots	[43]

[1] Apple ('Royal Gala') genes are from two SSH root libraries of water deficit-treated plants (manuscript submitted)

[2] This citation is a review of several different studies of water deficit-response in*Arabidopsis*.

6. GENES ASSOCIATED WITH WUE

Recent reports of genes associated with regulation of transpiration demonstrate the complexity of water use and transport, as well as the overlap in gene response to other stresses. *ESKIMO1*, which was originally associated with cold response, has recently been shown to affect both drought and salt responses in Arabidopsis. Insertional mutation lines inactivating *ESK1* had reduced transpiration rates and were only slightly less drought tolerant than wild type [45]. Furthermore, there was a reduction in biomass when mutant plants were grown without stress, suggesting that alterations in WUE were reducing both transpiration and CO_2 exchange; biomass differences between WT and *esk1* lines were negligible under stress. Another example of pleotrophic gene effects on plant physiology was reported in a study by Masle et al. [46]. A leucine-rich repeat receptor kinase (*ERECTA*) previously associated with inflorescence development was also shown to regulate transpiration in different races of Arabidopsis. The gene is implicated in epidermal cell expansion, cell-cell contact and mesophyll cell proliferation, but its relationship to reduced transpiration may be linked to differences in stomatal density between lines. Interestingly, there was no compensation in biomass for the reduction in E, indicating that as with WUE and $\Delta^{13}C$, there are workable strategies for increasing WUE without sacrificing carbon assimilation under normal growth conditions.

Using a suppression subtractive hybridization (SSH) approach to drought-responsive gene isolation in a commercial apple, 'Royal Gala', we identified numerous genes commonly found to be WD responsive in other plants. We also identified several up-regulated genes unique to apple roots (Table 2; manuscript submitted). Some of these genes may reflect the role of roots in nutrient transport during stress, including a copper chaperone and a high affinity nitrate transporter that is a putative Arabidopsis NRT2.4 homolog.

Table 2. Genes uniquely up-regulated in WD-treated apple tissues[1]

Genes Up-regulated in 'Royal Gala'		
Roots	**Bark**	**Leaves**
BYPASS1	Anthocyanin reductase	Auxin/Aluminum-induced protein
Serine acetyltransferase	GAST1-like gene	Proteasome maturation factor
High affinity nitrate transporter	Asparagine synthetase	Asparagine synthetase
NPR1	Chloroplast membrane protein Tic40	Glyoxylate aminotransferase

[1] These genes have not yet been reported as up-regulated in response to WD in these tissues of other plants.

7. GENES ASSOCIATED WITH DROUGHT AVOIDANCE AND ESCAPE

An example of the rapid evolution of a drought escape mechanism (early flowering) was demonstrated in a population of *Brassica napa* subjected to a multiyear drought [47]. Comparison of seeds collected from individual plants before the drought with those obtained from the drought-affected population indicated a significant earlier onset of flowering in the latter population. This observation was further expanded in a study of quantitave trait loci (QTLs) in maize associated with flowering time (*Vgt1*) where a *cis*-acting region upstream of a transcription factor was shown to be the link between a QTL and the early flowering trait [48]. The authors speculate that natural genetic variations in flowering time enabled the selection of maize lines adapted to a range of latitudes and growing seasons, including drought tolerance. Another study of natural variation in ecotypes of Arabidopsis [49] found a strong positive genetic correlation (rG = 0.98) between $\delta^{13}C$ (drought avoidance) and flowering time (drought escape). They also observed compelling evidence for pleiotropy in lines varying in FLOWERING LOCUS C, suggesting that correlated evolution of $\delta^{13}C$ and flowering time could be partly explained by coordinated allele fixation altering both traits.

In alfalfa a gene encoding a zinc-finger motif is expressed in roots [50, 51]. The protein encoded by*Alfin1* binds DNA at a specific *cis*-elemtent and is proposed to be a root growth regulator, as transgenic lines overexpressing the gene show enhanced root growth under both normal and high salt conditions [52]. These same transgenic lines were significantly more salt tolerant, and presumably more drought tolerant (high salt concentrations in the media decrease water uptake), althought drought tolerance per se was not measured.

8. GENES ASSOCIATED WITH DROUGHT TOLERANCE AND RESISTANCE

Studies of specific genes associated with dehydration responses have been conducted in a number of plants, and roles for many of these genes have been correlated with specific morphological or physiological traits known to be involved in drought resistance. Abscisic acid (ABA) signaling and stomatal function are correlated with WUE and drought resistance, so it is not surprising that several genes involved in ABA perception and stomata opening/closing respond to severe dehydration. Two calcium-dependent protein kinases from Arabidopsis have been implicated in slow-type anion channel activation [53]. In the double mutants, ABA and Ca^{2+}-induced stomatal closing were impaired, but not completely. These genes may contribute to a rapid Ca^{2+}-reactive response resulting in stomatal closure, as opposed to the slower Ca^{2+}-programmed response which maintains long-term stomatal closure. Similar studies have also implicated a G protein-coupled receptor, GCR1, in ABA signaling perception in guard cells and during seed germination [54]. To examine stomatal function in more detail, Klein et al. [55] used a T-DNA insertion disrupting At*MRP5*, an ABC transporter in Arabidopsis. The mutants had reduced transpiration rates and showed increased water use efficiency. In a similar study using T-DNA insertion disruption of At*MRP4* (another type of ABC transporter), Klein et al. [56] found the mutant lines to be more drought susceptible. In a recent study of the effects of overexpressing or silencing early response to dehydration (*ERD15*) in Arabidopsis, Kariola et al. [57] observed decreased drought tolerance in the overexpressing lines, whereas the silenced lines were hypersensitive to ABA and showed enhanced tolerance to drought. This study also suggests a negative role for *ERD15* in mediating stress-related ABA signaling.

The cuticle is an important barrier to moisture loss in plants, therefore genes associated with cuticle synthesis and turnover are expected to contribute to plant water status. A recent study in alfalfa demonstrated that increased wax production activated by a putative TF (transcription factor) [*WXP1*] also enhanced drought tolerance in transgenic plants [58]. Similar studies with other transcription associated factors have also demonstrated a correlation between increased wax synthesis and drought tolerance [59, 60].

Constitutive expression of a barley Group III LEA (late embryogenesis abundant) protein placed in wheat under control of the maize *ubi1* promoter resulted in improved water use efficiency and higher total dry mass in the majority of transgenic lines [61]. Overexpressing a small molecular weight heat shock protein (*HSP17.6*) conferred drought tolerance in Arabidopsis transgenic lines [62]. The authors also demonstrated that *HSP17.6* had chaperone-like activity and could protect citrate synthase from chemical denaturation. Taken together these studies emphasize the different mechanisms so far discovered that affect water use and drought resistance in plants and indicate that different genes in the same pathway may be used by plants to control WD responses.

9. REGULATION OF PATHWAYS/SIGNALING NETWORKS ASSOCIATED WITH DEHYDRATION

Most studies of dehydration responsive signaling pathways implicate ABA directly in altering specific gene expression [63]. In fact genes that respond to ABA usually have multiple copies of an ABA response element (ABRE) or a combination of an ABRE with other motifs such as Myb, Myc or coupling elements [for example, 64]. A second pathway involves drought response element binding (DREB) proteins, particularly *DREB2*-encoding genes, and may also involve ABA indirectly [65]. A recent report on the isolation of a fourth CBF (*CBF4*) from Arabidopsis suggests that this TF only responds to drought, in contrast to observations reported for CBFs1-3 which respond to both cold and drought stress [66].

Compelling evidence indicates that the ABA pathway likely involves Ca^{2+} signal transduction as an early step and important relay system for dehydration responses. ABA can also increase reactive oxygen species through higher levels of H_2O_2 [67, reviewed in 68]. Other studies have suggested stress-responsive pathways that operate through osmotic sensing independently of ABA [69]. An osmotic sensor similar to bacterial two-component receptors has been identified in Arabidopsis [70]. The gene was able to complement several mutations in yeast osmosensors and activated the *HOG1*response pathway through a mitogen-activated protein kinase. No doubt other signaling components, both ABA-dependent and independent, will be identified in the near future.

Numerous studies support the existence of extensive cross-talk between plant hormone signaling pathways [71-74]. It is therefore expected that both the salicylic acid and Jasmonic acid (JA)/ethylene pathways indirectly influence the expression of genes that respond to drought. Along these lines an Arabidopsis mutant (*rcd1*) belonging to the ADP-(ribosyl)transferase domain-containing subfamily of the WWE family exhibits reduced sensitivity to ABA, ethylene and Me-JA, suggesting that it acts at an integrative node in hormonal signaling regulating different stress-responsive genes [75]. *rcd1* is just one example of many where one gene participates in multiple pathways.

10. STUDIES IN *MALUS SIEVERSII*

Genetic polymorphisms from twenty populations of *M. sieversii* in Xinjiang, China were analyzed with RAPD markers to assess genetic diversity [76]. Based on the bands generated with 42 randomly chosen primers, variation within a population (83.1%) was higher than among populations (16.9%). The authors conclude that *M. sieversii* is a rich source of genetic diversity.

Evaluation of six *Malus* species using a variety of morphological and physiological traits led to the conclusion that *M. toringoides* and *M. sieversii* were the most drought tolerant of those analyszed [77]. Measurement of root parameters indicated that *M. sieversii* root surface area decreased in response to drought to ~25% of the well-watered control in contrast to *M. toringoides* roots which decreased to ~ 43% of the control. On the other hand, root surface activity (absorption) declined the least in *M.toringoides* (17%) compared to *seiversii* (33%). When data from all the measurements were taken into consideration, *M. toringoides* and *M. seiversii* were the top two most drought resistant species.

A study of the contribution of rootstock source to drought resistance was conducted using 'Gale Gala' apple scions grafted onto *Malus sieversii* or *Malus hupehensis* roots [78]. Differential responses of the grafted material to drought stress were determined by a number of physiological and morphological traits. Typical reductions in biomass, growth rate and leaf area are observed under drought conditions, but *M. sieversii* showed smaller reductions in these traits during drought treatment than *M. hupehensis*. Furthermore, a larger increase in whole plant WUE was measured in grafts on *M. sieversii* rootstocks compared to *M. hupehensis*.

Problem Statement: Many of our agronomically important fruit trees are derived from a rather narrow genetic base. To provide methods for enhancing apple germplasm resistance to drought and other dehydrative abiotic stresses it is imperative that we identify those genes that contribute to drought survival. Once these genes are identified and characterized they can be used in marker assisted selection strategies or altered by genetic engineering.

Application Area: The origin of the domesticated apple is believed to be in Central Asia via the silk route through Kazakhstan [79]. The predominant species contributing to the domestication of the modern apple is believed to be *Malus sieversii* which is thought to be the progenitor of *M.* × *domestica*and a possible source of resistance alleles lost during domestication [80]. A significant secondary contributor to the genetics of current apple varieties is the European crabapple, *Malus sylvestris* [79] which may also possess important resistant genes lost in the modern varieties.

Several studies of *M. sieversii* material collected from geographically and climatologically different sites in Kazakhstan have concluded that significant genetic diversity can be captured in small-sized sub-populations of these site collections [81, 82]. In a subsequent study of the Kazakhstan collection, Richards et al. [83] concluded that differentiation in genetic diversity was greater among individual families than among sites and that gene diversity and allelic richness varied significantly among collection sites. The use of this material to study drought responses in apple at the morphological and genetic levels without the complication of grafted rootstocks provides the cornerstone of our approach to identifying novel drought resistant mechanisms or factors contributing to drought susceptibility.

Figure 1. *Malus sieversii* collection sites in Kazakhstan [after 80].

Table 3. Example of phenotypic diversity of select lines from *M. sieversii* Kazakhstan populations

Individual Seedling ID	Site[a]	Rainfall[b]	$\delta^{13}C$[c]	Fire Blight	Scab[d]	Juiciness	Surface Russet
GMAL3975.k	6	250 mm	-27.09 ppt	R	R	dry	0
GMAL3685.e	6	250 mm	-29.30 ppt	R	S	dry	10%
GMAL3623.f	9	450 mm	-26.33 ppt	S	S	moderate	1%
GMAL4455	4	800 mm	-26.30 ppt	R	R	medium	20%

[a] Geographical location of populations in Kazakhstan.

[b] Annual rainfall; source Forsline et al. (2003)

[c] ppt = parts per thousand; differences of 0.5 ppt are significant.

[d] R = resistant, S = susceptible

Research Course: In order to isolate and characterize genes responding to drought from a commercially important cultivar as a standard for comparison, we used SSH on cDNA prepared from 'Royal Gala' subjected to a moderate-severe drought. Genes identified by this method were further characterized for their expression in various tissues under drought treatment or in fully watered controls.

To begin analyzing *M. sieversii* lines for drought responsiveness, we first surveyed core populations of individuals collected from xeric site 6 and later, xeric site 9 for WUE using stable carbon isotope composition ($\delta\ ^{13}C$) to identify individuals with better WUE. Morphological features, e.g. leaf area, leaf length, stomatal density and number of leaves per current year's branch length were evaluated [84]. Individuals showing $\delta\ ^{13}C$ extreme values were then clonally propagated for simulated drought experiments where photosynthesis and stomatal conductance were determined and roots, bark and leaves were collected for gene expression analysis. We duplicated these studies using 'Royal Gala', a relatively drought tolerant variety for comparison [85].

Methods Used: We used standard methods for the morphological and physiological measurements. For the drought experiments, young trees (~1 m tall) of 'Royal Gala' propagated by shoot proliferation were grown for several weeks in a controlled environment with standard light and temperature conditions [44]. A simulated moderate-severe drought were imposed by withholding water until the pots reached 40% of full saturation and maintained for 2 weeks at this level after which the trees were sampled. A parallel control group was grown under the same conditions, but watered to full capacity every other day. Samples from roots, bark and leaves (fully expanded) were taken and quickly immersed in liquid N_2. Roots were washed for 5 min in room temperature tap water, blotted dry and placed in liquid N_2. Bark was removed by scraping the outer layers (down to the xylem) directly into liquid nitrogen. All samples were stored at -80°C until use. Bark was lyophilized prior to storage at -80°C.

Total RNA was isolated, cDNA prepared and SSH performed using the protocol reported by Bassett et al. [86] for peach. For gene analysis, we designed primers for several genes shown to be associated with dehydration responsiveness in apple [44; manuscript submitted]. Each primer pair was quality tested and used to prime RT-qPCR reactions in order to quantitate gene expression in different tissues. The qPCR reactions were conducted using a kit containing all reagents (Life Technologies, Applied Biosystems, Grand Island, NY) and the reaction parameters were as follows: 95°C 5 min, followed by 35 cycles of 95°C 1 min, 60-65°C 1 min, 72°C 1 min and a final extension of 72°C for 10 min. Primers for a translation elongation factor (TEF2) was used as an internal control for the qPCR experiments [87]. The relative standard curve method was used to analyze the data.

10.1. Status and Results

Analysis of 'Royal Gala' response to a simulated moderate-severe drought

Clonally replicated individuals growing on their own roots were generated for the suppression subtractive hybridization experiments. We identified several hundred different genes up-regulated or down-regulated in roots, bark and leaves when DNA from drought-treated tissues served as 'tester' and DNA from well watered controls acted as 'driver' (10-fold higher amount). Some genes unique to our experiments that increased in response to drought are shown in Table 2. A number of drought-responsive genes were common to genes isolated from other plants systems, both dicot and monocot. Figures 2-4 show the relative expression of some of the common and unique genes in apple roots, bark and leaves.

Most of the genes examined regardless of tissue origin showed around a two-fold difference between watered and water-deficit treaments. A few genes were substanitally up-regulated in response to drought. including the auxin-induced gene from leaves (8-fold increase) and asparagine synthase from bark (4-fold increase). A few genes in roots were also significantly up-regulated in response to drought treatment, one of which was NPR1 (3-fold increase; manuscript submitted). From this information, including the expression of genes not shown here, we have generated a list of potential up-regulated genes responding to a relatively long term drought that can be used to determine if there is any correlation of expression in *M. sieversii* lines with high and low WUE values.

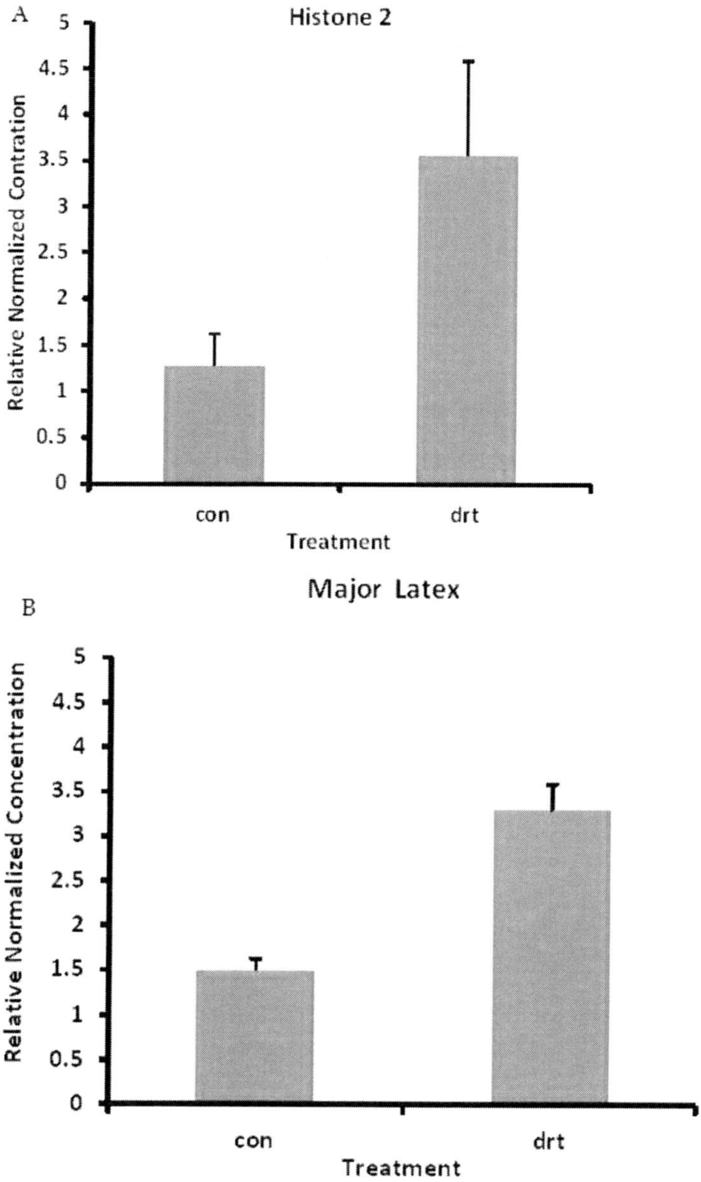

Figure 2. Relative expression of root genes up-regulated by drought treatment. Experiments were conducted as described in Methods. Test transcripts were normalized against TEF2. con: well watered controls; drt: treated to 40% saturation for two weeks. A: relative expression of Histone H2b gene; B: relative expression of the Major Latex protein gene.

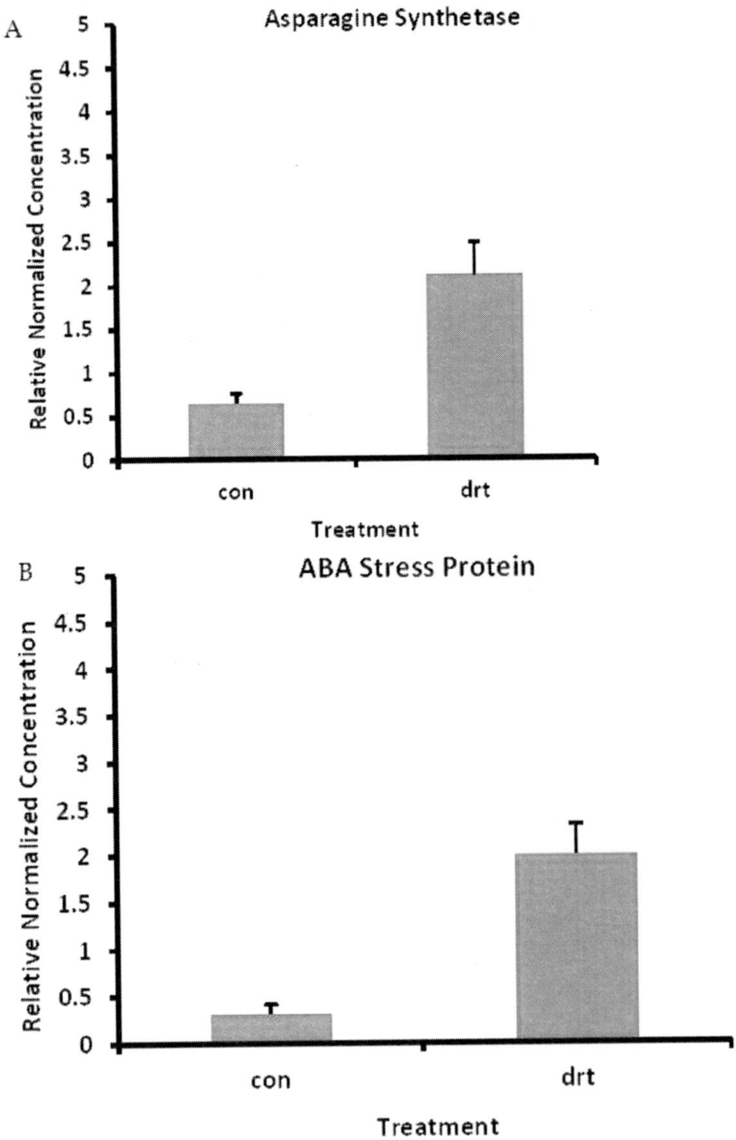

Figure 3. Relative expression of bark genes up-regulated by drought treatment. Experiments were conducted as described in Methods. Test transcripts were normalized against Actin. con: well watered controls; drt: treated to 40% saturation for two weeks. A: relative expression of asparagine synthetase gene; B: relative expression of an ABA stress protein gene.

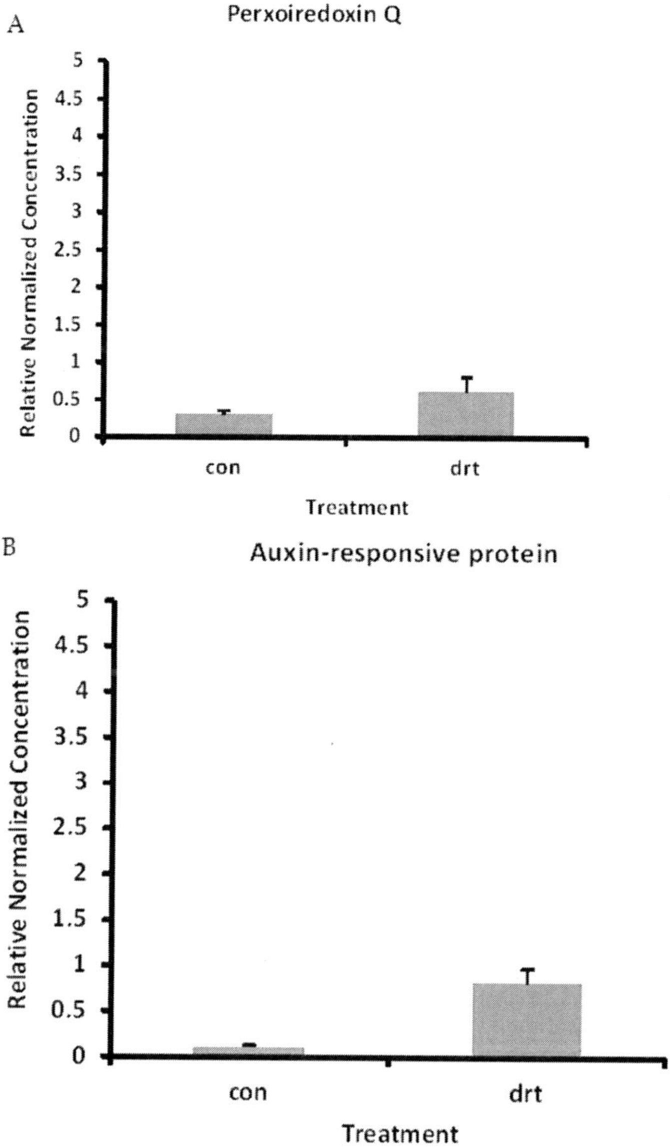

Figure 4. Relative expression of leaf genes up-regulated by drought treatment. Experiments were conducted as described in Methods. Test transcripts were normalized against Actin. con: well watered controls; drt: treated to 40% saturation for two weeks. A: relative expression of a peroxiredoxin gene; B: relative expression of an auxin-responsive/ripening protein gene.

10.2. Promoter Comparison Of Nrt2.4 From Apple, Peach And Arabidopsis

A high affinity nitrate transporter gene from the 'Royal Gala' root SSH library (see Table 2) was shown by RT-qPCR to be elevated in roots and bark in response to drought treatment (manuscript submitted). Approximately 700 bp upstream of the ATG start codon was obtained from the genomic sequence of 'Golden Delicious' (Genome Database for Rosaceae; http://www.rosaceae.org/). Several *cis*-elements associated with tissue specificity or stress response were identified. To identify elements preserved during evolution, promoter regions from peach NRT2.4 and an Arabidopsis AtNRT2.4 gene were analyzed for comparison (Figure 5). All three genes contained consensus, well defined TATA boxes within 80-100 bp of the translation start. A number of MYB and MYC binding sites were identified in similar positions in all three promoters. The peach promoter was missing a root-specific element seen in both MdNRT2.4 and AtNRT2.4. Interestingly, both the apple and peach promoters contained a drought responsive element on the reverse strand. Overall the elements identified in the apple NRT2.4 promoter are consistent with the expression analysis results.

Screening the *M. sieversii* population at site 6 in Kazakhstan

A core diversity population of *M. sieversii* trees (34 individuals representing 14 sibling groups) collected from Kazakhstan site 6 and maintained as seedlings at the Geneva, NY ARS Plant Genetic Resources Unit, was sampled for stable carbon isotope discrimination to select individuals with extreme values compared to 'Royal Gala'. The results are shown in Figure 6. Two individuals from each end of the WUE spectrum were chosen for further characterization.

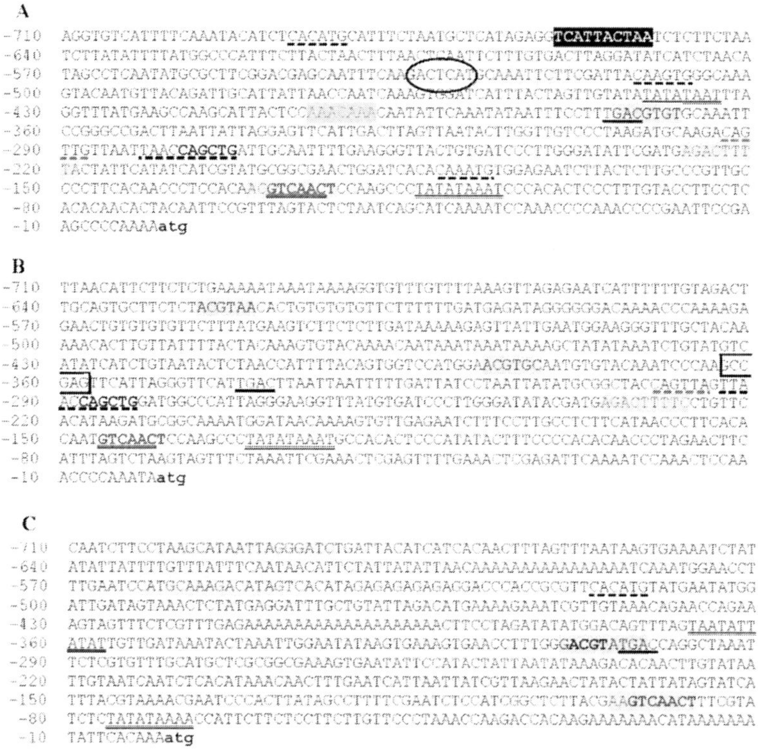

Figure 5. Figure 5. Comparison of NRT2.4 promoter regions from apple and Arabidopsis. Only the first 700 bases upstream of the translation start (**atg**) are shown. *Cis*-elements were identified by PLACE [88] and PLANTCare [89]. A: Promoter region from apple MdNRT2.4; B: Promoter region from peach genome; C: Promoter region from Arabidopsis AtNRT2.4. **AAACAAA:** anaerobic induction [90]; TGACG: WRKY stress responsive binding element [91]; **A^{A}/$_{C}$GTCA** and **^{G}/$_{A}$GACTTTTC:** bZIP and NF-B binding sites, respectively [92]; **CAAGCATGCTTCTTGC:** consensus root-spscific element [93]; **TATA box:** PolII binding; **TCATTACTAA:** wound-inducible element [94]; ACGTG/AT: ABRE core element [95]. An element of unknown function in the Arabidopsis NRT2.4 gene promoter (**GTCAACT**) is also present in the MdNRT2.4 and peach promoters. A *cis*-element for hypoosmolarity-responsiveness [96] is identified by an oval. Dashed underlines indicate a MYB (WAACCA) binding site [97] and MYC core sequences (CANNTG); fuschia-underlined MYC element on the negative strand in A and B (CAACTG) is associated with drought response [73]. A CBF/DREB element [98, 99] is boxed in A and B.

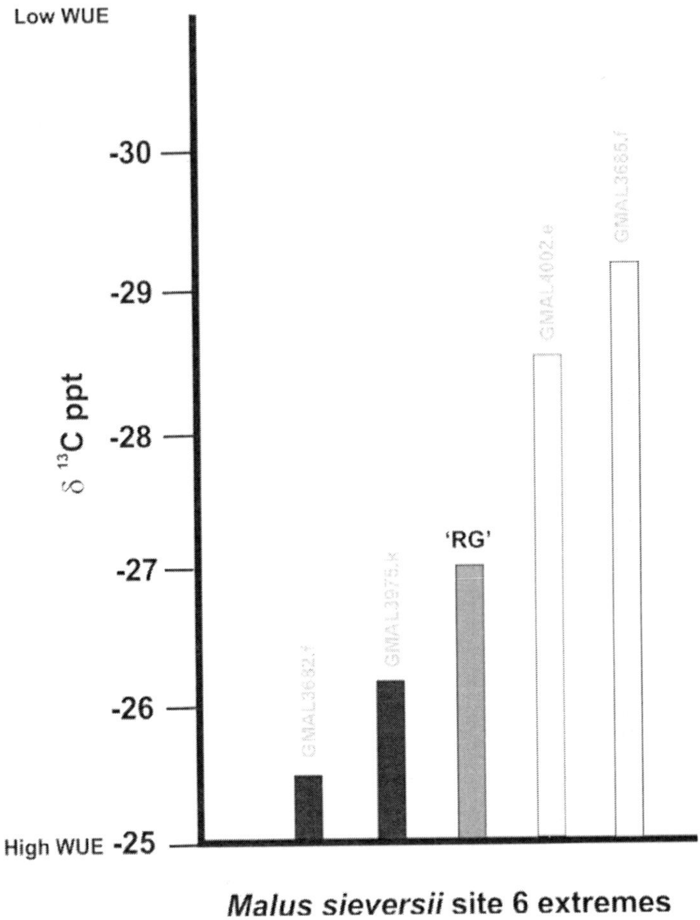

Figure 6. Stable carbon isotope analysis of select *M. sieversii* individuals from the site 6 population Measurements were made from branches representing current year's growth. Collection of samples was made from dormant trees for three years and averaged. RG: 'Royal Gala' standard.

M. sieversii lines GMAL4002.e and GMAL3975.k were propagated by shoot proliferation techniques to obtain a number of clonal individuals on their own roots. At the same time, 'Royal Gala' was propagated as a standard for comparison. Individuals from each line (4-6 trees per treatment) were acclimated under controlled conditions of light, water, fertilizer and temperature. Water was withheld from half of each group, while the other half received sufficient water to saturate the pot. The water-deficit trees reached 40% of pot saturation in about 7-10 days, at which time the individual plant and a comparable control were tagged. At the end of two weeks of treatment, the plants were individually sampled. Out of six GMAL4002.e plants, four began to show signs of severe wilting within a few days after reaching 40% as illustrated in Figure 7. This is consistent with WUE measurements which indicated that this particular line was not adept at maintaining healthy water status under the water deficit regime. Well watered GMAL4002.e controls showed no signs of wilting throughout the experiment. In contrast, GMAL3975.k with a WUE close to that of 'Royal Gala' showed no signs of wilting under the well-watered regime or water deficit stress (Figure 6). These results indicate that WUE can be used to identify apple lines that are drought sensitive, as well as drought tolerant.

Leaf size and number have been shown to respond negatively to drought, resulting in longer intervals between newly initiated leaves and smaller sizes, all features designed to reduce transpiration to conserve water. Leaf morphological features and stomatal characteristics were examined in the site 6 subpopulation [84]. GMAL3683.o had the smallest leaves by all traits measured, whereas GMAL3687.d and GMAL3989.f had the largest leaves by area. Within the GMAL3683 sibling group, GMAL3683.o leaf area (8.3 cm^2) clearly segregated from the other three members (average = 20.3 cm^2).

Stomatal density has also been linked to drought tolerance and sensitivity. Therefore, we also examined stomata size and density in the site 6 *M. sieversii* population. GMAL3691.m had the largest stomates, whereas GMAL3684.a had the smallest. GMAL3689.n had the highest density of stomates (58 per 0.09 mm^2); GMAL3685.e had the lowest density (23 per 0.09 mm^2). There was no correlation between leaf and stomate features in the *M. sieversii* site 6 population, nor was there a correlation between the leaf and stomate extreme values and the ^{13}C extreme values. In other words, although leaf and stomate physical characteristics might contribute to the drought response of *M. sieversii*individuals from site 6, another mechanism(s) appears to better explain the WUE data.

Figure 7. 'Royal Gala' and different *M. sieversii* genotypes under severe water deficit conditions. Delta ^{13}C values are included under the cultivar/genotype name. A difference of 0.50 ppt is considered significant. Note the wilting observed with GMAL4002.e compared to 'Royal Gala' under identical SD conditions. Also note GMAL3975.k and 'Royal Gala' respond similarly to the SD. 13C = $\delta 13$C

Further Research: The overall goal of this project is to link drought responsiveness and/or WUE to specific apple genes. To this end we are interested in candidate genes that are either up-regulated in response to drought (drought 'defensive' genes) or down-regulated (drought survival 'assisting' genes). The latter can be used to identify genetic alterations that could hamper the physiological state attained by up-regulated genes and therefore to be avoided in breeding programs. Up-regulated genes are of interest because of their obvious association with drought resistance. We have developed primers for many of the drought up-regulated genes identified in 'Royal Gala' and have tested them against individual lines of *M. sieversii*. Experiments to quantify their expression in the roots, bark and leaves of the site 6 lines representing WUE extremes are currently underway. We are also replicating additional*M. sieversii* lines on their own roots for simulated drought experiments like the ones shown in Figure 7 to provide physiological, morphological and molecular biological data to detect associations to drought resistance or susceptibility. Since the *M. sieversii* site 6 and 9 populations have undergone rapid adaptation to the xerophytic environments they currently occupy, it seems likely that alterations in patterns of expression could account for their survival. To this end we are planning experiments to determine the kinetics of expression of candidate genes over a broader time period in select *M. sieversii* lines. Finally, we plan to examine promoter elements for single nucleotide polymorphisms or other alterations that might influence expression and to identify changes in methylation that might contribute to differences in expression between the more drought resistant lines and those that are more sensitive. The information generated from these experiments can be used in breeding programs to select drought resistant progeny using marker assisted selection. In addition the use of genetic engineering of select genes is another potentially successful approach to obtaining new varieites with improved drought resistance or enhanced WUE. With increasing competition between agriculture and urban populations for fresh water and with climate change which predicts increasing episodes of intense drought periods worldwide, there is a critical need for the development of crop varieties with more efficient use of water and the ability to survive longer drought periods. Since apples are a good source of nutrition and can be conveniently stored and shipped, they are a logical target for improvement.

REFERENCES

1. Bohnert HJ, Nelson DE. and Jensen RG. Adaptations to environmental stresses. The Plant Cell 1995; 7(7):1099-111.

2. Ingram J. and Bartels D. The molecular basis of dehydration tolerance in plants. annual review of Plant physiology and plant molecular biology 1996; 47:377-403.

3. Street NR, Skogstrom O., Sjodin A., Tucker J., Rodriguez-Acosta M., Nilsson P., Jansson S. and Taylor G. (2006) The genetics and genomics of the drought response in *populus*. plant journal 2006; 48(3):321-341.

4. Bogeat-Triboulot M-B., Brosché M., Renaut J., Jouve L., Le Thiec D., Fayyaz P., Vinocur B., Witters E., Laukens K., Teichmann T., Altman A., Hausman J-F., Polle A., Kangasjärvi J. and Dreyer E. Changes of gene expression, protein profiles, ecophysiology and growth performance in*Populus euphratica*, a poplar growing in arid regions. Plant Physiology 2007; 143(2):876-892.

5. Maseda PH. and Fernández RJ. Stay wet or else: three ways in which plants can adjust hydraulically to their environment. Journal of Experimental Botany 2006; 57(15):3963-3977.

6. Blum A. Drought resistance, water-use efficiency, and yield potential: are they compatible, dissonant, or mutually exclusive? Crop & Pasture Science (formerly Australian Journal of Agricultural Research) 2005; 56(11):1159-1168.

7. Levitt J. Responses of plants to environmental stresses. New York: Academic Press; 1972.

8. Dudley SA. Differing selection on plant physiological traits in response to environmental water availability: A test of adaptive hypotheses. Evolution 1996; 50(1):92-102.

9. Dudley SA. The response to differing selection on plant physiological traits: Evidence for local adaptation. Evolution 1996; 50(1):103-110.

10. Monclus R., Dreyer E., Villar M., Delmotte FM., Delay D., Petit J-M., Barbaroux C., Le Thiec D., Bréchet C. and Brignolas F. Impact of drought on productivity and water use efficiency in 29 genotypes of *Populus deltoides* × *Populus nigra*. New Phytologist 2006; 169(4):765-777.

11. Kramer PJ. and Boyer JS. Water Relations of Plants and Soils. New York: Academic Press; 1995.

12. Farquhar GD., Ehleringer JR. and Hubick KT. Carbon isotope discrimination and photosynthesis. Annual Review of Plant Physiology and Plant Molecular Biology 1989; 40:503-537.

13. Richards RA., Rebetzke GJ, Condon AG. and van Herwaarden AF. Breeding opportunnities for increasing the efficiency of water use and crop yield in temperate cereals. Crop Science 2002; 42(1):111-121.

14. Condon AG. and Richards RA. Broad sense heritability and genotype x environment interaction for carbon isotope discrimination in field-grown wheat. Crop & Pasture Science (formerly the Australian Journal of Agricultural Research) 1992; 43(5):921-934.

15. Condon AG., Richards RA., Rebetzke GJ. and Farquhar GD. Improving intrinsic water-use efficiency and crop yield. Crop Science 2002; 42(1):122-131.

16. Rebetzke GJ., Condon AG., Richards RA. and Farquhar GD. Selection for reduced carbon-isotope discrimination increases aerial biomass and grain yeild of rainfed bread wheat. Crop Science 2002; 42(3):739-745.

17. Ackerly DD., Susley SA., Sultan SE., Schmitt J., Coleman JS., Linder CR., Sandquist DR., Geber MA., Evans AS., Dawson TE. and Lechowicz MJ. The Evolution of Plant Ecophysiological Traits: Recent Advances and Future Directions. BioScience 2000; 50 (2) 979-995.

18. Knight CA., Vogel H., Kroymann J., Shumate A., Witsenboer H. and Mitchell-Olds T. Expression profiling and local adaptation of *Boechera holboellii* populations for water used efficiency across a naturally occurring water stress gradient. Molecular Ecology 2006; 15(5):1229-1237.

19. Sandquist DR. and Ehleringer JR. Intraspecific variation of leaf pubescence and drought adaptation in *Encelia farinosa* associated with contrasting desert environments. New Phytologist 1997; 135(4):635-644.

20. Sandquist DR. and Ehleringer JR. Intraspecific variation of drought adaptation in brittlebush: Leaf pubescence and timing of leaf loss vary with rainfall. Oecologia 1998; 113(2):162-169.

21. Aspelmeier S. and Leuschner C. Genotypic variation in drought response of silver birch (*Betula pendula* Roth): leaf water status and carbon gain. Tree Physiology 2004; 24(5):517-528.

22. Bonal D., Sabatier D., Montpied P., Tremeaux D. and Guehl J-M. Interspecific variability of delta 13C among trees in rainforests of French Guiana: functional groups and canopy interactions. Oecologia 2000; 124(3):454-468.

23. Brendel O., Pot D., Plomion C., Rozanberg P. and Guehl J-M. Genetic parameters and QTL analysis of delta 13C and ring width in maritime pine. Plant, Cell & Environment 2002; 25(8):945-953.

24. Ponton S., Dupouey J-L., Bréda N., Feuillat F., Bodenes C. and Dreyer E. Carbon isotope discrinimation and wood anatomy variation in mixed stands of *Quercus robur* and *Quercus petraea*. Plant, Cell & Environment 2001; 24(8):861-868.

25. Aspelmeier S. and Leuschner C. Genotypic variation in drought response of silver birch (*Betula pendula* Roth): leaf and root morphology and carbon partitioning. Trees 2006; 20(1):42-52.

26. Mäkelä A., Berninger F. and Hari P. Optimal control of gas exchange during drought: theoretical analysis. Annals of Botany 1996; 77(5):461-467.

27. Marron N., Villar M., Dreyer E., Delay D., Boudouresque E., Petit J-M., Delmotte FM., Guehl JM. and Brignolas F. Diversity of leaf traits related to productivity in 31 *Populus deltoides* × *Populus niger* clones. Tree Physiology 2005; 25(4):425-435.

28. Monclus R, Dreyer E, Delmotte FM, Villar M, Delay D, Boudouresque E, Petit J-M, Marron N, Bréchet C and Brignolas F. Productivity, leaf traits and carbon isotope discrimination in 29 *Populus deltoides* × *P. nigra* clones. New Phytolologist 2005; 167(1):53-62.

29. Dennis FG., Jr. The history of fruit thinning. Plant Growth Regulation 2000; 31(1):1-16.

30. Glenn DM., Scorza R. and Okie WR. Genetic and environmental effects on water use efficiency in peach. Journal of the American Society for Horticultural Science 131(2):290-294.

31. Farquhar GD., Cernusak LA. and Barnes B. Heavy water fractionation during transpiration. Plant Physiology 2007;143(1):11-18.

32. Barbour MM., Fischer RA., Sayre KD. and Farquhar GD. Oxygen isotope ratio of leaf and grain material correlates with stomatal conductance and grain yield in irrigated wheat. Australian Journal of Plant Physiology 2000; 27:625-237.

33. Barbour MM., Fischer RA., Sayre KD., Farquhar GD.. Oxygen isotope ratio of leaf and grain material correlates with stomatal conductance and grain yield in irrigated wheat. Australian Journal of Plant Physiology 2000; 27: 625–637.

34. Houde M., Belcaid M., Ouellet F., Danyluk J., Monroy AF., Dryanova A., Gulick P., Bergeron A., Laroche A., Links MG., MacCarthy L., Crosby WL. and Sarhan F. Wheat EST resources for functional genomics of abiotic stress. BioMed Central Genomics 2006; 7:149 http://www.biomedcentral.com/1471-2164/7/149.

35. Ozturk ZN., Talamé V., Deyholos M., Michalowski CB., Galbraith DW., Gozukirmizi N., Tuberosa R. and Bohnert HJ. (2002) Monitoring large-scale changes in transcript abundance in drought- and salt-stressed barley. Plant Molecular Biology 2002; 48(5-6):551-573.

36. Pratt LH., Liang C., Shah M., Sun F., Wang H., Reid SP., Gingle AR., Paterson AH., Wing R., Dean R., Klein R., Nguyen HT., Ma H-M., Zhao X., Morishige DT., Mullet JE. and Cordonnier-Pratt M-M. Sorghum expressed sequence tags identify signature genes for drought, pathogenesis, and skotomorphogenesis from a milestone set of 16,801 unique transcripts. Plant Physiology 2005; 139(2):869-884.

37. Reddy AR., Ramakrishna W., Sekhar AC., Ithal N., Babu PR., Bonaldo MF., Soares MB. and Bennetzen JL. Novel genes are enriched in normalized cDNA libraries from drought-stressed seedlings of rice (*Oryza sativa* L. subsp. indica cv. Nagina 22). Genome 2002; 45(1):204-211.

38. Seki M., Narusaka M., Ishida J., Nanjo T., Fujita M., Oono Y., Kamiya A., Nakajima M., Enju A., Sakurai T., Satou M., Akiuama K., Taji T., Yamaguchi-Shinozaki K., Carninci P., Kawai J., Hayashizaki Y. and Shinozaki K. Monitoring the expression profiles of 7000 Arabidopsis

genes under drought, cold, and high-saliinity stresses using a full-length cDNA microarray. Plant Journal 2002; 31(3):279-292.

39. Kilian J., Whitehead D., Horak J., Wanke D., Weinl S., Batistic O., D'Angelo C., Bornberg-Bauer E., Kudla J. and Harter K. The AtGenExpress global stress expression data set: protocols, evaluation and model data analysis of UV-B light, drought and cold stress responses. Plant Journal 2007; 55(3):526-542

40. Bray EA. Classification of genes differentially expressed during water-deficit stress in *Arabidopsis thaliana*: an analysis using microarray and differential expression data. Annals of Botany 2002; 89(7):803-811.

41. Plomion C., Lalanne C., Claverol S., Meddour H., Kohler A., Bogeat-Triboulot M-B., Barre A., Le Provost G., Dumazet H., Jacob D., Bastien C., Dreyer E., de Daruvar A., Guehl J-M., Schmitter J-M., Martin F. and Bonneu M. (2006) Mapping the proteome of poplar and application to the discovery of drought-stress responsive proteins. Proteomics 2006; 6(24):6509-6527.

42. Boominathan P., Shukla R., Kumar A., Manna D., Negi D., Verma PK. and Chattopadhyay D. Long term transcript accumulation during the development of dehydration adaptation in *Cicer arietinum*. Plant Physiology 2004; 135(3):1608-1620.

43. Jia J., Fu J., Zheng J., Zhou X., Huai J., Wang J., Wang M., Zhang Y., Chen X., Zhang J., Zhao J., Su Z., Lu Y. and Wang G. Annotation and expression profile analysis of 2073 full-length cDNAs from stress-induced maize (Zea mays L.) seedlings. Plant Journal 2006; 48(5):710-727.

44. Wisniewski M., Bassett C., Norelli J., Macarisin D., Artlip T., Gasic K., Korban S. Expressed sequence tag analysis of the response of apple (*Malus x domestica*) to low temperature and water deficit, Physiologia Plantarum 2008; 133:298-317.

45. Bouchabke-Coussa O., Quashie ML., Seoane-Redondo J., Fortabat MN., Gery C., Yu A., Linderme D., Trouverie J., Granier F., Téoulé E. and Durand-Tardif M. *ESKIMO1* is a key gene involved in water economy as well as cold acclimation and salt tolerance. BioMed Central, Plant Biology 2008

46. Masle J., Gilmore SR. and Farquhar GD. The *ERECTA* gene regulates plant transpiration efficiency in Arabidopsis. Nature 2005; 436(Aug 11):866-870.

47. Franks SJ., Sim S. and Weis AE. Rapid evoluton of flowering time by an annual plant in response to a climate fluctuation. Proceedings of the National Academy of Sciences USA 2007; 104 (4):1278-1282.

48. Salvi S., Sponza G., Morgante M., Tomes D., Niu X., Fangler KA., Meeley R., Ananiev EV., Svitashev S., Bruggemann E., Li B., Hainey CF., Radovic S., Zaina G., Rafalski J-A., Tingey SV., Miao G-H., Phillips RL. and Tuberosa R. Conserved Noncoding genomic Sequences Associated with a Flowering-time Quantitative Trait Locus in Maize. Proceedings of the National Academy of Sciences USA 2007;104(27) 11376-11381.

49. Mckay JK., Richards JH. and Mitchell-Olds T. Genetics of Drought Adaptation in Arabidopsis thaliana: 1. Pleiotropy Contributes to Genetic Correlations Among Ecological Traits. Molecular Ecology; 2003 12(5) 1137-1151.

50. Winicov I. cDNA Encoding Putative Zinc Finger Motifs from Salt-Tolerant Alfalfa (*Medicago sativa* L.) Cells. Plant Physiology 1993; 102(2) 681-682.

51. Bastola DR., Pethe VV. and Winicov I. Alfin1, A Novel Zinc-Finger Protein in Alfalfa Roots That Binds to Promoter Elements in the Salt-Inducible MsPRP2 Gene. Plant Molecular Biology 1998; 38(6):1123-1135.

52. Winicov I. Alfin1 Transcription Factor Overespression Enhances Plant Root Growth Under Normal and Saline Conditions and Improves Salt Tolerance in Alfalfa. Planta 2000; 210 (3) 416-422.

53. Mori IC., Murata Y., Yang Y., Munemasa S., Wang Y-F., Andreoli S., Tiriac H., Alonso JM., Harper JF., Ecker JR., Kwak JM. and Schroeder JI. CDPKs CPK6 and CPK3 function in ABA regulation of guard cell S-type anion- and Ca2+-permeable channels and stomatal closure. Public Library of Science Biology 2006; 4:e327.

54. Pandey S. and Assmann SM. The Arabidopsis putative G protein-coupled receptor GCR1 interacts with the G protein α subunit GPA1

and regulates abscisic acid signaling. The Plant Cell 2003; 16(6):1616-1632.

55. Klein M., Perfus-Barbeoch L., Frelet A., Gaedeke N., Reinhardt D., Mueller-Roeber B., Martinoia E. and Forestier C. The plant multidrug resistance ABC transporter At*MRP5* is involved in guard cell hormonal signalling and water use. Plant Journal 2003; 33(1):119-129.

56. Klein M., Geisler M., Suh AJ., Kolukisaoglu Ü., Azevedo L., Plaza S., Curtis MD., Richter A., Weder B., Schulz B. and Martinoia E. Disruption of At*MRP4*, a guard cell plasma membrane ABCC-type ABC transporter, leads to deregulation of stomatal opening and increased drought susceptibility. Plant Journal 2004; 39(2):219-236.

57. Kariola T., Brader G., Helenius E., Li J., Heino P. and Palva ET. EARLY RESPONSIVE TO DEHYDRATION 15, a negative regulator of abscisic acid responses in Arabidopsis. Plant Physiology 2006; 142(4):1559-1573.

58. Zhang J-Y., Broedkling CD., Blancaflor EB., Sledge MK., Sumner LW. and Wang Z-Y. Overexpression of *WXP1*, a putative *Medicago truncatula*AP2 domain-containing transcription factor gene, increases cuticular wax accumulation and enhances drought tolerance in transgenic alfalfa (*Medicago sativa*). Plant Journal 2005; 42(5):689-707.

59. Aharoni A., Dixit S., Jetter R., Thoenes E., van Arkel G. and Pereira A. The SHINE clade of AP2 domain transcription factors activates wax biosynthesis, alters cuticle properties, and confers drought tolerance when overexpressed in Arabidopsis. The Plant Cell 2004; 16(9):2463-2480.

60. Broun P., Poindexter P., Osborne E., Jiang C-Z. and Riechmann JL. *WIN1*, a transcriptional activator of epidermal wax accumulation in Arabidopsis. Proceedings of the. National. Academy of Sciences USA 2004; 101(13):4706-4711.

61. Sivamani E., Bahieldin A., Wraith JM., Al-Niemi T., Syer WE., Ho DT-H. and Qu R. Improved biomass productivity and water use efficiency under water deficit conditions in transgenic wheat constitutively expressing the barley *HVA1* gene. Plant Science 2000; 155(2):1-9.

62. Sun W., Bernard C., van de Cotte B., Van Montagu M. and Vergruggen N. At-HSP17.6A, encoding a small heat-shock protein in Arabidopsis,

can enhance osmotolerance upon overexpression. Plant Journal 2001; 27(5):407-415.

63. Busk PK. and Pagès M. Regulation of abscisic acid-induced transcription. Plant Molecular Biology 1998; 37(3):425-435.

64. Shen QJ., Casaretto JA., Zhang P. and Ho T-HD. Functional definition of ABA-response complexes: the promoter units necessary and sufficient for ABA induction of gene expression in barley (*Hordeum vulgare* L.). Plant Molecular Biology 2004; 54(1):111-124.

65. Knight H., Zarka DG., Okamoto H., Thomashow MF. and Knight MR. Abscisic acid induces CBF gene transription and subsequent induction of cold-regulated genes via the CRT promoter element. Plant Physiology 2004; 135(3):1710-1717.

66. Haake V., Cook D., Riechmann JL., Pineda O., Thomashow MF. and Zhang JZ. Transcription factor CBF4 is a regulator of drought adaptation in Arabidopsis. Plant Physiology 2002; 130(2):639-648.

67. Miao YC., Song C-P., Dong FC. and Wang XC. ABA-induced hydrogen peroxide generation in guard cells of *Vicia faba*. Acta Phytophysiolgia Sinica 2000; 26(1):53-58.

68. Mittler R., Vanderauwera S., Gollery M. and Van Breusegem F. Reactive oxygen gene network of plants. Trends in Plant Science 2004; 9(10):490-498.

69. Assmann SM., Snyder JA. and Lee Y-RJ. ABA-deficient (aba1) and ABA-insensitive (abi1-1, abi2-1) mutants of Arabidopsis have a wild-type stomatal response to humidity. Plant Cell & Environment 2000; 23(4):387-395.

70. Urao T., Yakubov B., Satoh R., Yamaguchi-Shinozaki K., Seki M., Hirayama T. and Shinozaki K. A transmembrane hybrid-type histidine kinase in Arabidopsis functions as an osmosensor. The Plant Cell 1999; 11(9):1743-1754.

71. Christmann A., Moes D., Himmelbach A., Yang Y., Tang Y. and Grill E. Integration of abscisic acid signalling into plant responses. Plant Biology 2006; 8(3):314-325.

72. Mishra NS., Tuteja R. and Tuteja N. Signaling through MAP kinase networks in plants. Archives of Biochemistry and Biophysics 2006; 452(1):55-68.

73. Yamaguchi-Shinozaki K. and Shinozaki K. The plant hormone abscisic acid mediates the drought-induced expression but not the seed-specific expression of rd22, a gene responsive to dehydration stress in Arabidopsis thaliana. Molecular Genetics and Genomes (formerly called Molecular and General Genetics) 1993; 238(1-2):17-25.

74. Zhang T., Liu Y., Yang T., Zhang L., Xu S., Xue L. and An L. Diverse signals converge at MAPK cascades in plants. Plant Physiology and Biochemistry 2006; 44(5-6):274-278.

75. Ahlfors R., Lang S., Overmyer K., Jaspers P., Brosché M., Tauriainen A., Kollist H., Tuominen H., Belles-Boix E., Piippo M., Inzé D., Palva ET. and Kangasjärvi J. Arabidopsis RADICAL-INDUCED CELL DEATH1 belongs to the WWE protein-protein interaction domain protein family and modulates abscisic acid, ethylene and methyl jasmonate responses. The Plant Cell 2004; 16(7):1925-1937.

76. Yan G., Long H., Song W. and Chen R. Genetic polymorphism of *Malus sieversii* populations in Xinjiang, China. Genetic Resources and Crop Evolution 2008

77. Yang J., Yang E. and Yang H. A study on drought resistance of genus Malus seedlings (In Chinese). Acta Agriculturae Boreali-Sinica 1996; 11(2):81-86.

78. Liu B-H., Cheng L., Ma F-W., Liang D. and Zou Y-J. Influence of rootstock on drought response in young 'Gale Gala' apple (*Malus domestica*Borkh.) trees. Journal of the Science of Food and Agriculture 2012.

79. Cornille A., Gladieux P., Smulders MJM., Roldán-Ruiz I., Laurens F. et al. New insight into the history of domesticated apple: secondary contribution of the European wild apple to the gnome of cultivates varieties. Public Library of Science, Genetics, 2012.

80. Forsline PL., Aldwinckle HS., Dickson EE., Luby JJ. and Hokanson SC. Collection, Maintenance, Characterization and Utilization of Wild

Apples of Central Asia. In: Janick J. (ed.) Horticultural Reviews. John Wiley & Sons, Hoboken, NJ 2003; 29:1-62.

81. Volk GM., Richards CM., Reilley AA., Henk AD., Forsline PL. and Aldwinckle HS. Ex Situ Conservation of Vegetatively Propagated Species: Development of a Seed-based Core Collection for Malus sieversii. Journal of the American Society for Horticultural Society 2005; 130(2):203-210.

82. Richards CM., Volk GM., Reeves PA., Reilley AA., Forsline PL. and Aldwinckle HS. Selection of Stratified Core Sets Representing Wild Apple (Malus sieversii). Journal of the American Society for Horticultural Science 2009; 134(2):228-235.

83. Richards CM., Volk GM., Reilley AA., Henk AD., Lockwood DR., Reeves PA. and Forsline PL. Genetic Diversity and Population Structure in Malus sieversii, a Wild Progenitor Species of Domesticated Apple. Tree Genetics & Genomes 2009; 5(2):339-347.

84. Bassett CL., Glenn DM., Forsline PL., Wisniewski ME. and Farrell Jr. RE. Characterizing water use efficiency and water deficit responses in apple (*Malus* × *domestica* Borkh. and *Malus sieversii* Ledeb.) M. Roem. HortScience 2011; 46(8):1079-1084.

85. Nemeskéri E., Sárdi É,, Szabó T. and Nyéki J. Ecological drought resistance and adaptability of apple varieties. International Journal of Horticultural Science 2010; 16(1):113-122.

86. Bassett CL., Wisniewski ME., Artlip TA., Norelli JL., Renaut J. and Farrell, Jr RE. Global analysis of genes regulated by low temperature and photoperiod in peach bark. Journal of the American Society for Horticultural Science 2006; 131(4):551-563.

87. Tong Z., Gao Z., Wang F., Zhou J. and Zhang Z. Selection of reliable reference genes for gene expression studies in peach using real-time PCR. BioMed Central Molecular Biology 2009

88. Higo K., Ugawa Y., Iwamoto M. and Korenaga T. Plant cis-acting regulatory DNA elements (PLACE) database:1999. Nucleic Acids Research 1999; 27(1):297-300.

89. Lescot M., Déhais P.,Thijs G., Marchal K., Moreau Y., Van de Peer Y., Rouzé P. and Rombauts S. PlantCARE, a database of plant cis-acting

regulatory elements and a portal to tools for in silico analysis of promoter sequences. Nucleic Acids Research 2002; 30(1):325-327.

90. Mohanty B., Krishnan SP., Swarup S. and Bajic VB. Detection and preliminary analysis of motifs in promoters of anaerobically induced genes of different plant species. Annals of Botany 2005; 96(4):669-681.

91. Maleck K., Levine A., Eulgem T., Morgan A., Schmid J., Lawton KA., Dangl JL. and Dietrich RA. The transcriptome of Arabidopsis thaliana during systemic acquired resistance. Nature Genetics 2000; 26(4):403-410.

92. Lebel E., Heifetz P., Thorne L., Uknes S., Ryals J. and Ward E. Functional analysis of regulatory sequences controlling PR-1 gene expression in Arabidopsis. Plant Journal 1998; 16(2):223-233.

93. Kobayashi T., Nakayama Y., Itai RN., Nakanishi H., Yoshihara T., Mori S. and Nishizawa NK. Identification of novel cis-acting elements, IDE1 and IDE2, of the barley ISD2 gene promoter conferring iron-deficiency-inducible, root-specific expression in heterogeneous tobacco plants. Plant Journal 2003; 36(6):780-793.

94. Pastuglia M., Roby D., Dumas C. and Cock JM. Rapid induction by wounding and bacterial infection of an S gene family receptor-like kinase in Brassica oleracea. The Plant Cell 1997; 9(1):1-13.

95. Shen Q. and Ho T-HD. Functional dissection of an abscisic acid (ABA)-inducible gene reveals two independent ABA-responsive complexes each containing a G-box and novel cis-acting element. The Plant Cell 1995; 7(3):295-307.

96. Satoh R., Nakashima K., Seki M., Shinozaki K., and Yamaguchi-Shinozaki K. ACTCAT, a novel cis-acting element for proline- and hypoosmolarity-responsive expression of the ProDH gene encoding proline dehydrogenase in Arabidopsis. Plant Physiology 2002; 130(2):709-719.

97. Abe H., Yamaguchi-Shinozaki K., Urao T., Iwasaki T., Hosokawa D. and Shinozaki K. Role of Arabidopsis MYC and MYB homologs in drought- and abscisic acid-regulated gene expression. The Plant Cell 1997; 9(10):1859-1868.

98. Yamaguchi-Shinozaki K. and Shinozaki K. A novel cis-acting element in an Arabidopsis gene is involved in responsiveness to drought, low-temperature, or high-salt stress. The Plant Cell 1994; 6(2):251-264.

99. Baker SS., Wilhelm KS. and Thomashow MF. The 5N region of Arabidopsis thaliana cor15a has cis-acting elements that confer cold-, drought- and ABA-regulated gene expression. Plant Molecular Biology 1994; 24(5):701-713.

Index